Fish Protein Concentrate

International Nutrition Policy Series

1. *Morinda: An Economic Analysis of Malnutrition Among Young Children in Rural India,* F. James Levinson, 1974

2. *The Children of Santa María Cauqué: A Prospective Field Study of Health and Growth,* Leonardo J. Mata, 1977

3. *Fish Protein Concentrate: Panacea for Protein Malnutrition?,* E. R. Pariser, Mitchel B. Wallerstein, Christopher J. Corkery, and Norman L. Brown, 1978

All correspondence about submission of manuscripts and allied matters should be sent to International Nutrition Policy Series, Mass. Inst. of Tech., 20A-222, 18 Vassar Street, Cambridge, MA 02139. Correspondence about ordering books or subscribing to the series should be sent to The MIT Press, 28 Carleton Street, Cambridge, MA 02142.

Fish Protein Concentrate: Panacea for Protein Malnutrition?

Ernst R. Pariser
Christopher J. Corkery
Mitchel B. Wallerstein
Norman L. Brown

The MIT Press
Cambridge, Massachusetts, and London, England

Funding for this study was provided in part by a grant to M.I.T. from the U.S. Agency for International Development and in part by a grant to M.I.T. from the Rockefeller Foundation. The volume represents solely the views of the authors and not necessarily those of either sponsor.

This book was set in IBM Composer Theme by To The Lighthouse Press, printed and bound by Halliday Lithograph Corp., in the United States of America

Library of Congress Cataloging in Publication Data
Main entry under title:

Fish protein concentrate.

(International nutrition policy series)
1. Proteins in human nutrition. 2. Fish protein concentrate. 3. Nutrition policy.
I. Pariser, Ernst R. II. Series.
TX553.P7F56 641.1'2 77-28112
ISBN 0-262-16069-2

To Geulah, Zadikah
We offer this book with these hopes for the hungry: that the mistakes described here will not be repeated; that the lessons learned will be heeded.

Contents

Series Foreword

The magnitude and complexity of the malnutrition problem and its relationship to national development have been increasingly recognized by governments in low-income and industrialized countries, bilateral and international assistance agencies, foundations, and the academic community. Although large-scale program success to date has been limited, a capacity, backed up by field experience, is being developed to address the problem and to carry out effective nutrition planning.

Unfortunately, there has been no means of systematically making available to concerned persons information on alternative approaches to nutrition planning and programming and on actual nutrition intervention programs that have been attempted.

To meet this need, the MIT International Nutrition Planning and Policy Program and the Cornell University Program on International Nutrition and Development Policy jointly initiated the *International Nutrition Policy Series.* This continuing series will focus on two main categories of monographs: (1) national or regional planning studies, utilizing tools of the natural and social sciences to identify and refine appropriate nutrition policy instruments; and (2) reports and reviews of specific nutrition intervention programs, providing accounts of actual projects, information on projected costs and effectiveness, and evaluation of the impact and significance of such programs.

The editors and editorial board bring to the task both considerable professional competence in the natural and social sciences and a great deal of international experience. The board has set for itself the highest standards of excellence to assure a series that will be of professional quality and of value in the application and teaching of international nutrition policy.

Foreword

To the nutrition field the late 1960s might be characterized as the era of the technological fix. Oilseed isolates, single cell protein, leaf protein concentrate, fish protein concentrate, and synthetic amino acids (particularly lysine) were all widely publicized at one time or another as technological panaceas to the world's malnutrition problem. Yet within several years the potential if not the viability of each was brought into question.

During this latter period of skepticism we were jointly teaching a course in nutrition planning at MIT. In 1974, when that time came in the semester to direct attention to the new technologies, we decided to use the occasion to step back and examine the lessons learned—good and bad—from experience with the new protein sources. The most appropriate approach to an in-depth class analysis seemed to be concentration on one of the technological fixes and, in looking at the alternatives, none of the panaceas put forth had been fraught with more problems, political pressures, frustrations, and public notoriety than fish protein concentrate (FPC). Moreover, unlike most attempts at nutrition intervention, the FPC story had a beginning, a middle, and, in most people's opinion, an end; presumably something could be learned for the benefit of future endeavors.

It was only natural to turn to Ray Pariser to provide a case study for the class discussion. As a food chemist, he had been identified with FPC development through most of its history and had personally been involved or was familiar with many of the political, economic, technical, and managerial controversies. The resulting lecture presented by Ray (and assisted by his long-time colleague, Norman Brown, then of the National Academy of Sciences) was titled "FPC: The One that Got Away."

For the students, this scholarly, thoughtful, honest appraisal was a highlight of the course. Subsequently the class concluded that the presentation deserved a wider audience, and Ray Pariser was persuaded to undertake the necessary work that resulted in this volume.

Assisting Pariser in this effort was Norman Brown, also a food chemist with many years' experience in FPC development; Mitchel Wallerstein, a political scientist working in the field of food and nutrition policy; and Christopher Corkery, a writer with an overview of nutrition planning. Their aim was to discuss and analyze the FPC history and glean from the experience the relevant lessons for those interested in international nutrition. In carrying out the project, the group studied what probably is the largest body of information ever assembled on the subject. They also interviewed or corresponded with most of the central characters in the FPC story.

The conclusion of this analysis is that fish protein concentrate is "a sophisticated and capital-intensive technological solution in search of a problem to which it can be applied." According to the authors, FPC, like other technological fixes, was doomed from the outset by "conceptional myopia."

This fascinating case study breaks new ground in nutrition literature and provides important insights on the need for building better bridges between the technical and policy communities.

Alan Berg and F. James Levinson

Introduction

Fish protein concentrate (FPC) is a stable, wholesome protein supplement of high nutritive value and low caloric content prepared from whole fish.[1] Developed in an era of enthusiasm for the potential contributions of technology to human welfare, FPC and its technology inspired high hopes. The most prevalent concept of malnutrition during this period was that of protein deficiency which had its origins in the pioneering work of Cicely Williams in Africa in the 1930s. Williams described the kwashiorkor she treated there as a protein deficiency disease, and this description gradually was applied to malnutrition in widely varying locales, with little realization that the disease as it existed in Africa differed from the marasmic type of malnutrition encountered there and in other parts of the world. This kind of broad definition lent itself to the traditional medical model of disease causation and cure, which posits that each disease is caused by a specific agent and that the cure for each disease lies in combating this agent effectively. The idea of protein deficiency thus led naturally to the idea of protein cure. The effective use of skim milk in kwashiorkor cure in Africa corroborated this association.

The division between well- and malnourished populations soon became known as the "protein gap." At the United Nations, the Protein Advisory Group (PAG) was set up in 1955 to provide knowledge and direction in the development of new protein foods for use in combating malnutrition through UN agency aid projects. Thus the notion that the "protein gap" could be bridged by developments in food technology gained great appeal, creating a momentum in research and development efforts that proved to be inexorable throughout the 1960s. Within this context expectations for FPC greatly exceeded and ultimately distorted its real potential as a factor in the alleviation of malnutrition.

Today the protein gap concept has been significantly modified, not because the gap itself has been narrowed but because malnutrition is no longer considered as a single factor deficiency. The term "world food crisis" has replaced "protein gap" in popular parlance, connoting both greater urgency and much more complexity. Thus, at a time when new and economical sources of food are being sought and when the politics and ethics of food aid are being seriously questioned, we believe that there is much to be learned from the controversial, well-documented, and relatively recent history of the various attempts to develop FPC as a new food component. Our objective in this examination of the history of FPC development is to understand why, given these many international attempts and high expectations, this product has had so little impact on its anticipated goal of alleviating malnutrition.

We believe that the insights to be gained from such an analysis are important and timely because, despite diminished faith in technological solutions and a broadened concept of the nature of malnutrition, food products like FPC continue to be developed. Like FPC, they are based on new technologies, they are products previously unknown to the consumer, and most important, they are intended to address the increasing shortage of traditional food sources in both developed and developing countries.

In considering the FPC experience and the technological, political, bureaucratic, industrial, and cultural interactions that formed the matrix of success and failure for this product, it will be useful for the reader to bear in mind the following considerations:

1. The FPC technology was developed because the idea, the technical capacity, and the will to create the product existed. This is the environment of basic research. Unfortunately in this case, as in many other examples of technological development, the convergence of these factors constituted a technological imperative that became the driving force, that is, the technology was developed because it could be developed. In the ensuing rush of activity critical questions concerning the practicality and applicability of the product were overlooked or ignored. For FPC the technological imperative was made even more compelling by the fact that the product was envisioned as serving a vital humanitarian purpose, namely, that of feeding the hungry.
2. Though this humanitarian goal was admirable, its meaning was too often obscured in the course of political and technological disputes. Nutrition activists, politicians, regulatory agencies, and private industry came into conflict in a manner most often detrimental to the implementation of FPC as a nutrition supplement. Differing goals among developers and conflicting priorities within and between institutions severely constrained the development of FPC, particularly in terms of the following issues:

a. Inadequate consideration was given to the social characteristics, cultural patterns, and nutritional needs of the group or groups for whom the product was intended. Although many of the cultural connotations of fish are unique, the fact remains that every traditional food exists within a cultural context, and any nutrition intervention that involves modification of food habits must know and respect this context.
b. Selection of product characteristics was directed toward the development of a universally acceptable product, namely, a greyish white, odorless, tasteless powder. This orientation was self-defeating since it imposed technical and economic constraints that profoundly affected the cost and applicability of the product.
3. The production of FPC is a high-technology enterprise, requiring massive inputs of capital, energy, and expertise. All of these inputs are characteristically lacking in the developing world. The attempt to transfer FPC technology demonstrates that technological solutions that seem plausible in a developed context are less advantageous and usually inappropriate when transferred to a developing context.
4. Given the high cost of FPC technology, the issue of how the production and consumption of the final product would be financed became critical. In an international food system where food is acquired according to the economic status of the consumer or consuming nation, the cost of producing a supplement such as FPC cannot be met by the impoverished groups for whom the product is intended, that is, through the marketplace. In the course of FPC development, however, the question of who would pay for the product once it had moved beyond the development stage was never effectively addressed.
5. The principal development of FPC occurred within the public sector due to the high risk and capitalization required to successfully develop a new technology. However, this notion of FPC as a public good was not uniformly accepted within the private sector. In cases of public sector development where government subsidy and political support were diminished or withdrawn, development efforts were curtailed or stopped entirely.
6. The concept of a nutritional supplement usually implies a product that is used in addition to something else. In the case of FPC the product supplements the nutritional quality of a food but cannot, by itself, increase significantly the quantity and caloric value of the total diet of a target population that has, to begin with, less than enough. In this context, the usefulness of protein supplementations that only marginally affect the nutritional status of a target group and do not affect in any way the group's absolute supply of food or its food purchasing power must be questioned. It is a consistent finding of this study that FPC did not positively affect either of these critical dimensions of malnutrition.

Over the past two decades FPC and its various production processes have been developed and refined in more than forty countries by many groups and individuals. Detailed information about the nature and extent of these endeavors is available, however, in only a limited number of these cases. We have also been confronted by the problem that FPC development efforts in the United States were larger, more heavily funded, and better documented than in any other comparable country or group of countries. In addition, certain major issues, pertaining to bureaucratic politics and to acceptable levels of filth in food, were raised almost exclusively in the U.S. context, although their resolution had major impacts on FPC development and acceptance elsewhere.

We have divided the presentation of the FPC case into a sectional format. Because many of the significant aspects of the story are related to the nature of the raw material in question, namely, fish, chapter 1 in part I contains a brief discussion of the history and cultural significance of fish as a food. Chapter 2 presents the most significant precedents for the development of FPC as well as a discussion of the early FPC development initiatives.

Part II of the case study is devoted exclusively to an examination of the history of policy issues pertaining to the promotion, regulation, and development of FPC in the United States. As the nation in which most of the FPC development initiatives occurred, the United States was the major source of publicity for the product as a potential weapon against protein malnutrition. It was, simultaneously, the scene of a long and often bitter bureaucratic struggle—between an agency interested in the promotion of FPC, the Bureau of Commercial Fisheries (BCF) and an agency concerned with the regulation of such products, the Food and Drug Administration (FDA)—whose outcome significantly influenced FPC development and use worldwide. Chapter 3 presents a comprehensive view of the history and politics of FPC development in the United States. In chapters 4 and 5 specific issues that dominated the course of this development are addressed: the issue of promoter versus regulator in the context of bureaucratic decision making and the evolution of acceptable standards for levels of filth in food.

In part III international FPC development efforts are addressed. We focus intensively on four country case studies: Canada, Morocco, Chile, and Sweden. Canada is examined first, in chapter 6, since the Halifax laboratory of the Canadian Fisheries Research Board was the site of some of the earliest and most critical contemporary research on the FPC production process, and since a major private sector development initiative was unsuccessfully attempted here. Morocco is discussed in chapter 7 as an example of an FPC program in a developing country under the aegis of international agencies where the coordination of multilateral, bilateral, and private aid became a major issue. This is followed in chapter 8 by the case of Chile, another FPC undertaking in the Third World, in which several different development efforts occurred. The Chilean case demonstrates many of the difficulties inherent in the integration of domestic development policy and foreign aid resources. Chapter 9 presents the unique but illustrative case of Sweden in which development of the product was undertaken almost exclusively through the private sector and where the only significant private sector capacity to produce FPC exists today. FPC development efforts in other countries for which at least limited information is available are discussed in chapter 10.

Part IV concludes the principal part of the study by presenting an analysis of the international FPC experience in terms of specific issues pertaining to the development of new foods or food components and the solution of malnutrition-related problems. The policy implications we have drawn from the case are also presented in this section. This is followed by annotated references that direct the interested reader to a wide variety of technical issues pertaining to the production of FPC and its related technologies.

Acknowledgments

We owe a heavy debt of gratitude to Alan Berg and Jim Levinson for their encouragement, support, criticisms, and friendship. Without them, this book would not have seen the light of day.

We thank Eugene Skolnikoff and Nevin Scrimshaw for their patient and important advice; it helped us immensely throughout this writing.

We are greatly indebted to a much larger number of people than we can possibly mention here, people who permitted us to interview them, who faithfully corresponded with us, allowed us to consult their files and, in particular, provided a critical mass of constructive comments. Particular thanks, however, must go to the following individuals whose involvement in the project improved the final manuscript substantially.

First and foremost, there is Jessie Janjigian, who read, re-read, and clarified the text so patiently and masterfully, and Amy Leiss who, over the two years, guided us over, around, and under many formidable obstacles. The others whom we would like to thank are listed here alphabetically:

J. S. Annan, Gerry Bernstein, Janet Brown, Bureau of Commercial Fisheries/NMFS Staff, James Carr, John Cordaro, Harold Crowther, Robert Ernst, Food and Drug Administration Staff, Roland Finch, Ted Greenwood, Peter Hakim, H. J. Hanham, John Haydon, I. Hornstein, George Knobl, Ezra Levin, Donald McKernan, Max Milner, A. C. Mosha, Harold Olcott, Ole Rostad, E. B. Ruth, Max Rutman, Julio Santa Maria, Harvey Sapolski, Virginia Sidwell, Arnold Singal, Donald Snyder, Georgio Solimano, Bruce Stillings, J. P. Tripplaar, Stewart Udall, and Edward Wenk.

Acronyms

AID Agency for International Development
BCF Bureau of Commercial Fisheries
BEPI Bureau d'Etudes et des Participations Industrielles
BOB Bureau of the Budget
CARE Cooperative for American Relief Everywhere
CFTRI Central Food Technological Research Institute
CNAE Companha Nacional De Alimentacão Escolar
CORFO Corporacion de Fomento de la Produccion Industrial
CPI Center for Industrial Protein Development
CPL Cardinal Proteins, Ltd.
CPP Concentrado De Proteina De Pescado
CRO Central Research Organization
CSIR Council for Scientific and Industrial Research
CSIRO Central Scientific and Industrial Research Organization
DANIDA Danish International Development Agency
DOC Department of Commerce
DOD Department of Defense
DOS Department of State
ECE Early Childhood Education
EDP Experiment and Demonstration Plant
FAO Food and Agriculture Organization of the United Nations
FDA Food and Drug Administration
FDD Food and Drug Directorate
FIRI Fishery Industry Research Institute
FRBC Fisheries Research Board of Canada

FTC Federal Trade Commission
GAO U.S. General Accounting Office
GOC Government of Chile
HEW Department of Health, Education, and Welfare
IAFMM International Association of Fish Meal Manufacturers
IFOP Chilean Institute for Fisheries Development
INCAP Instituto de Nutrición de Centro America y Panama
MARDI Malaysian Agricultural Research and Development Institute
MCH Maternal and Child Health Care
MOR Marine Oil Refiners of Africa, Ltd.
MSC Marine Sciences Council
NAS National Academy of Sciences
NCFPC National Center for FPC
NFMOA National Fish Meal and Oil Association
NHS National Health Service
NIH National Institutes of Health
NMFS National Marine Fisheries Service
NNRI National Nutrition Research Institute
NOAA National Oceanic and Atmospheric Administration
OMB U.S. Office of Management and Budget
PAG Protein Advisory Group
PAHO Pan American Health Organization
PASA Participating Agency Services Agreement
PCSIR Pakistan Council for Scientific and Industrial Research
SONAFAP Société Nationale de Farine Alimentaire de Poisson
SIDA Swedish International Development Authority
UBARI Union of Burma Applied Research Institute
UN United Nations
UNDP United Nations Development Program
UNICEF United Nations Children's Fund
UNIDO United Nations Industrial Development Organization
UNROD United Nations Relief Operation, Dacca
USAID U.S. Agency for International Development
USDA U.S. Department of Agriculture
USDI U.S. Department of the Interior
WFP World Food Program
WHO World Health Organization

PART I
THE HISTORY AND CULTURAL SIGNIFICANCE OF FISH AS FOOD

Chapter 1
The Use of Fish as Food

All available evidence indicates that fish has been used by humans as food since long before recorded history. Fish and indeed many other aquatic organisms have been hunted, gathered, and harvested wherever they have appeared, in every part of the globe. Because of the ubiquity of the resource and its established food value, the harvesting, trading, preserving, and preparing of fish have, for millenia, played an important role in practically all civilizations.

Fish—in the general sense of food of aquatic origin—is, however, the only universal staple that inhabits a medium hostile and dangerous to people. It is, therefore, small wonder that many and diverse beliefs and social attitudes developed around this resource, its captors, merchants, and consumers, endowing it with attributes and powers that grew to be more varied, colorful, and deeply ingrained than those associated with perhaps any other food. The rapid spoilage of fish and its attendant dangers to health also give aquatic creatures a special place in human affairs. The need to harvest fish and, simultaneously, to devise methods to avoid spoilage, has provoked the development of strict safety criteria, of precise eating habits and taboos, and, by trial and error no doubt, of techniques to prolong the time during which raw fish could be consumed with impunity.

Little seems to have changed since earliest times with regard to the desire of the consumer to be able to inspect raw fish carefully for freshness and safety. With the exception of Far Eastern fish sausages and fish sauces, fish is consumed almost exclusively qua fish, that is, in forms in which the original identity of the raw material is preserved. In many western countries, where

Japanese and other Far Eastern products are practically unknown, the only exceptions to this rule are products like fish sticks and gefilte fish, but these represent only a very small portion of the total amount of fish consumed.

With the exception of canning, conceived in the eighteenth century and introduced on an industrial scale in the nineteenth century, all the most widely used processes that are employed today are exceedingly old, dating back to prehistory, although equipment, handling, quality control, and so forth, have been greatly modified and improved since that time. The methods used today are drying, salting, smoking, pickling, icing, freezing, freeze drying, canning, and fermentation. Although fermentation methods are commonly used in the East today, western technology has not used them for the last two thousand years.

We believe it is noteworthy, then, that in spite of the ubiquity of the resource, the recognized nutritional value of fish, and the important economic contributions that fisheries have made in the development of many nations, fish is still consumed in the West as in ancient times and no basic technological advances (with the exception of canning) have been made in its preservation. These observations are significant in that they do not apply to the consumption or preservation of any other of the staple foods. (Very early in the history of food technology milk was processed and preserved in the form of butter or cheese; wheat was made into flour, dough, bread, and beer; meat into sausages, and so on). Efforts to produce FPC could thus be seen basically as attempts to break this pattern and to develop stable fish products as flexible dietary additives, analogous to the multipurpose ingredients and supplements produced from other staples such as milk, eggs, wheat, and meat.

FISH IN ANTIQUITY

Age and Variety of the Resource

The history of the use of fish as food in the Western world is older than any document we possess; it is an exceedingly varied and picturesque history, uninterrupted over millenia. The Chinese proverb, "Give a man a fish and he will live for a day, teach him to fish and he will have food for life," was as true for the Babylonians, the early Egyptians, or the Greeks as it was for the Chinese. There is a strong evidence that at the time of Urukagina in the third millenium B.C. an active and well-developed aquaculture was flourishing in Babylon.[1] In the time of Hamurabi, in the seventeenth century B.C., eighteen different species of fish were traded in the marketplace of Larsa, many more than in the average modern supermarket.[2] To judge from their numerous and precise wall paintings, the Egyptians had developed sophisticated methods of harvesting, dressing, and preserving fish, at least two thousand years before

Christ.[3] The library of Assurbanipal at Nineveh contained an enumeration and description of two hundred different kinds of fish.[4]

In the eighteenth century B.C. Homer describes the Hellespont as teeming with fish and "counts the sea's bounty in supplying fish as an element of prosperity. . . ." He also notes that "when Odysseus and his companions are in Thrinacia, they catch fish and fowl with bent hooks . . . and this seems to prove that these people had had practice and skill in catching fish."[5]

The historical age of fish as a food resource is matched by the variety of species that were consumed in antiquity. The variety and importance of fish in the time of Alexander the Great and his teacher Aristotle (300 B.C.) is attested to by the meticulous care that the great philosopher bestows upon the description of an astonishing number of aquatic animals. In his *History of Animals* Aristotle assembled an impressive body of observations, interpretations, and speculations on the biology, reproduction, physiology, anatomy, and nutritive attributes of all kinds of animals, including fish, which remained a standard model and unassailable thesaurus over the centuries.[6] Archestratos of Galen, a Greek poet and contemporary of Aristotle's, in his "gastronomic journey around the world" deals particularly with fish: where and when to catch the best, how to prepare them, and what to eat with them. Athenaeus himself, in Book VII of the *Deipnosophistas,* mentions over eighty different species of fish with recommendations on how to cook and serve them.

Every area around the Mediterranean and the Black Sea, the center of the classical Western world, was known by the seafood for which it was most famous: mussels from the Pontus, Megarian grey fish, Carystian sprat, Eretrian bream, crayfish from Scyra, tuna from Byzantium.

Social and Cultural Significance of Fish and the Fishing Trade

The sea was often regarded in antiquity as either the dwelling place or physical manifestation of a deity. Consequently the sea and its inhabitants have been characterized by beliefs and superstitions which have found expression in behavioral restrictions and taboos. Many of these survive in one form or another today.

Certain fish were sacred to the Greek sea god, Poseidon. They were kept in ponds and anyone who ate them could expect death, rather than good nutrition. The priests in the service of Poseidon were forbidden to utilize any creatures of the sea as food. On the island of Ortygia, holy to Artemis (protector of small animals as well as goddess of hunting) no fishing and no plowing of the land was allowed. Pythagoras held fish to be so sacred (perhaps because of their silent nature) that he never used any as food.[7] Legend has it that on one occasion he bought the entire catch of some fishermen and had it thrown back into the sea. According to an ancient legend related by Plutarch, fish are supposed to have moved the egg from which Aphrodite sprang to the shores of the Euphrates. As a reward, Zeus placed the fish among the stars. This is the origin of the fish symbol in that place of the Zodiac that corresponds to the season of the rains. In Egypt, ocean fish were so hated that the hieroglyphic for "hatred" was a pictogram of an ocean fish.[8]

The sacredness of the sea and of fish is also represented in both the Christian and Judaic traditions. Fish was frequently represented in the West as a substitute for wine in the scene of the Last Supper. The Roman writer Tertullian wrote that "we little fishes according to our Ichthys, Jesus Christ, are born in the water nor are we saved in any other way than by remaining in the water." In Jewish tradition also the pious were fish living in the water of the Torah.[9] On the other hand, the Jewish taboo against eating fish without scales and fins may originate from their resemblance to snakes.[10]

The variety of symbolic values, both positive and negative, attributed to fish is still evident in the attitudes of contemporary social and ethnic groups toward fish as food. The harvest, culture, trade, and consumption of fish, a food item that appears to have been so vital in the dietary and symbolic patterns of many early cultures, had profound effects not only on the beliefs but also on the social patterns of civilizations.

In Urukagina's time, we have important indications of the socioeconomic significance of aquaculture in Babylon. Fish hatcheries were usually installed by the poor for the enjoyment and use of wealthier citizens, who had ample means to protect their property and harvest. When, however, the poor built fish ponds for their own livelihood and food supply, and, with no means of redress, lost their fish to thieves, legislation was devised to protect and compensate the aggrieved parties.[11] Later, during the reign of Darius II (422 B.C.), the formulation of laws led to the establishment of a whole body of ordinances—the oldest of their kind—regulating the rental and service agreements concerning fish ponds and hatcheries.

Other instances illustrate the special position that fishermen occupied, and still occupy, in the fabric of civilizations. It is a special place, not only because the produce that fishermen offer for sale is much more rapidly and dangerously perishable than almost any other food, but also because fishing is conducted in, and dependent upon, a seemingly mysterious, unpredictable, and dangerous medium. Although, for instance, certain fishermen in pre-Christian Greece were exempt from taxation and important fish importers were even given full citizenship, fishermen have, by and large, never been treated well in western nations. Fishing was, and generally remains, an activity at the lower end of the social totem pole.

Because of easy perishability, the sale of fresh fish also became a particularly hazardous trade, requiring capital and perseverance in order to weather the vicissitudes of supply and demand. The small tradespeople who sold fish on the open market had to use all the tricks of the trade to be sure that they would not be left with a large amount of unsold and inedible merchandise by the end of the day. It is not surprising then that the trading of fish, also, especially fresh fish, was considered in many countries to be at the very bottom of the occupational scale.

The sale of fish in classical Athens was regulated by a "Food and Drug Administration,"[12] which insisted not only that fish above a certain size be handed over to the municipality—thus regulating in some way the fisheries resource—but also forbade fishmongers to sit down at their stalls, a measure designed to expedite their business. It was also mandatory that only fresh fish, wet out of the water, be offered for sale. Since it was extremely difficult in the warm Mediterranean climate to comply with this request, fish traders resorted to an ingenious method to circumvent these regulations. Men were hired to go from stall to stall, pick a fight with someone and suddenly faint in front of the displayed merchandise. The owner naturally felt obliged to revive the poor fellow and emptied a bucket of water (which fortunately stood close by) over the man—and his fish—thus combining charity with a useful act.

Economic Significance

In his discourse on politics Aristotle proclaims fisheries to be among the most important natural means of acquisition of property, comparable in significance, scale, and ubiquity, to war, hunting, and pillage.[13] In Egypt, for example, "the abundance of fish produced in the Nile was an invaluable provision of nature in a country which had neither pasture lands nor large herds of cattle, and where corn was the principal production."[14] From numerous pictorial and other records we know that fish from the Nile and Egyptian lakes was part of the staple basic diet of the poorer section of the population.[15] Fish appear on many of the earliest Egyptian wall paintings and bas-reliefs[16] that describe precisely not only the varied species of aquatic food that was brought to market but also the manner in which the fresh fish was prepared and sold. In the earliest representations, sale was by barter; later, at the time of the New Kingdom when weigh scales had been introduced, the goods were exchanged for copper or gold.[17]

The use of salt by Egyptians for the preservation of fish and many other perishables should also be noted.[18] The enormous significance of this most vital inorganic food component and preservative is self-evident and has remained unchallenged to this day.

In Greece and in its colonies an active fishing industry existed as early as Homeric times (twelfth century B.C.).[19] The professional fishermen who appear in the *Odyssey*[20] were selling their wares by the sea and probably further inland as well. Preserved fish was exported in great quantities to all parts of the empire from Miletus, the most powerful trading center in Anatolia. As a tribute to their commercial importance fish were depicted on coins minted in the important Greek fishing colonies of Abydos, Kyzikos, and Sinope. Furthermore, fish traders were offered tax exemption benefits if they established their businesses in Sybaris.[21] The great importance the Athenians attributed to the saltfish trade is reflected in the following statement from the Deipnosophists: "The Athenians set such store by salt fish that they actually enrolled the sons of C., the salt fish dealer, as citizens."[22]

The economic significance that fisheries had in Rome is, if anything, greater than in Greece. The Romans introduced large-scale aquaculture into Europe by building immense fish ponds which they stocked with rare species. Lucullus, for example, had a mountain pierced to introduce seawater into his fish vivaria. He managed them so well that at his death the fish in the ponds were sold for the equivalent of about $100,000.

Although the immense variety and rapidly expanding market structure of the Roman fishing trade during the heyday of the Roman empire is a striking feature of countless contemporary writings, no single example better illustrates the attitude of the Romans of Cicero's time than the craze for the surmullet or supermullet. The Romans were so desirous of specimens of this fish that they paid up to $240.00 for a six-pound fish.[23]

During the Middle Ages in Europe, fish, both by capture and culture, remained an important staple. However, with increasing sophistication in the development and processing of agricultural produce, the variety of fish and other aquatic animals customarily used as food began to decline, although the economic and political role that fish played in the affairs of Western, and indeed, other nations, retained its significance. Even a superficial examination of the history books will reveal that the merchant and fighting fleets of many nations are the direct descendants of the earliest fishermen's dugouts. The military and economic power of the Hanseatic League in Northern Europe is but one example; the development of the British navy financed by taxation of the fishing industry and the unsuccessful attempt by the British Government to introduce a second weekly "fyshe-day" (sic) are other illustrations. When a detailed socioeconomic history of fisheries for even one country is written, it will show the unique and far-reaching role this industry has played in the destinies of almost all the world's fishing nations.

Methods of Preservation

The purchase of fresh fish in an organized open marketplace evolved into an important, and still extant, ritual in which a series of visual, tactile, and olfactory tests are performed on the fish by the prospective customer, who searches for any subtle signs of incipient spoilage; hence, the age-old resistance of the average western consumer to the purchase of fish in any form other than its original one, that is, a form in which a complete and detailed inspection can be performed.

Whenever the fish harvest in an economy was sufficient to provide substance for organized trade, export, and/or sale to individuals, methods were devised to preserve the raw materials for safe storage and transportation over prolonged periods of time. The most prevalent and widespread methods of preservation developed in antiquity and still in use today include appropriate dressing of the fish, followed by either air- or sun-drying, pickling, smoking, and especially salting. All are methods that tend, in one way or another, to reduce microbial activity and the water content of the raw material, and/or to inactivate proteolytic and other enzymes.

Here it is appropriate to mention two fish preserving processing techniques that originated independently in the West over two thousand years ago. The first of these technologies was developed by a community of poor people on the coast of the Persian Gulf in the fourth century B.C. When Alexander the Great traveled on the eastern shore of the Gulf of Persia, he came upon a people appropriately called the "Ichthyophagi" or "Fish Eaters." Arrian, who gives an account of these travels, has the following notes in his records:

> Below the Gadrosians dwell the people called "The Fish Eaters" . . . thinking here to seize corn by force, Nearchus attacked the town, but the natives showed freely their flour, ground down from dried fish, but only a small quantity of corn and barley. Only a few of them fish, for few have proper boats or any skills; for the most part, it is the receding tide which leaves fish in pools which provide their catch. The more tender ones they eat raw, the larger and tougher ones they dry in the sun until quite sere and then pound them and make flour and a bread of them.[24]

This is probably the earliest documented instance in which a fish protein concentrate is used as an additive to another staple food. As far as we know, there is no evidence that the processing method of the Ichthyophagi was used elsewhere.

The other technology came from Greece and Rome. A vase unearthed not long ago in the ruins of Pompeii bears the inscription: *"Liquamen optimum siccatum ex officino Umbrici Agathopi."*[25] Freely translated, this means: "The best liquamen in the world is made in the factories of Umbricus Agathopus." This liquamen was a further development of the Greek fish sauce, garum, which had been known in Greece since early days and is even mentioned by Aeschylus.[26] Manufacture and trade of this food became very active throughout pre-Christian Greek history. The cookbook of Apicius, written in the first or second century A.D., describes the manufacture and use of liquamen.[27] Apicius records in detail the recipes of the upper- and middle-class families of Rome and indicates the method of manufacture of liquamen by a salt-controlled protein autolysis of various types of fish. Pliny tells us that liquamen was made from small fish that were otherwise useless and unsaleable. The most expensive type of liquamen was sold for about $40.00 for six pints.[28]

No longer known in Europe (except perhaps in a few places in Portugal), Liquamen was consumed in classical Rome as generally as its Southeast Asian counterpart is today, that is, as an important seasoning agent for many types of food. We do not know when Nuoc Mâm, Nam Pra, and all the other fish hydrolysates were first developed in the Far East, but there is a striking similarity between the contemporary methods used there to produce the various oriental fish sauces and the extinct technology for the manufacture of liquamen.

The two technologies are interesting for two reasons: (1) they anticipate the development of modern methods designed to utilize processed fish as a flexible food additive or supplement in a form in which its raw material identity is entirely lost; and (2) although apparently successfully used on a relatively large scale at the time, both techniques were forgotten by the Western world and have not been utilized there since antiquity. One thing is reasonably certain: neither of these processing methods is utilized in modern western fish technology, although millions of liters of fish sauces are consumed annually in Southeast Asia.

From what has just been said, however, it should not be concluded that the widespread and large-scale use of fermented and otherwise processed fish products is evidence of a general acceptance. Attitudes toward fish in that continent, as indeed, elsewhere, are as specific and widely varied as religious, cultural, and dietary patterns. Just as there are areas in India and adjacent countries, for example, where fish is accepted as a basic staple, there are equally large regions where fish has not been and is not being consumed at all. This is especially true for those communities where categoric dietary restrictions are in force for religious reasons, such as those that exist among the strict vegetarian followers of Buddhism and Hinduism. For these people, the consumption of fish, as well as meat, is excluded since it would violate their beliefs in the sanctity of life.

Although fish avoidance in India is most widely based upon philosophical and religious concepts, the development of this attitude, according to Simoons,[29] may also have been influenced by other factors, two of which should be mentioned here. Some Indo-European tribes, especially in areas of nomadic tradition, consider fish as dirty creatures and therefore fishing and fish-eating as low class. The nineteenth-century Kafirs, for instance, who lived in the Hindu Kush region of Afghanistan, are known to have detested fish[30] although the rivers of Kafiristan offered an abundance of this food; the Kafirs "shuddered at the idea of eating fish much as we would react to the thought of eating rats." According to Schomberg, the ordinary people of Gilgit (linguistically related to the Kafirs and also living in India's Northwest) will consume fish when it is available, but the upper classes refuse to do so because of the food's offensive smell.[31]

Another factor in people's attitude toward fish and one of great antiquity in India may have been the widespread conviction that fish living in sacred waters are themselves holy creatures[32] and therefore must not be injured or caught. In fact, these animals are, to this day, carefully fed and cared for.

Fish avoidance patterns are found all over Asia and are also well documented in Africa. In the latter case attitudes range from simple indifference to fish in Ethiopia as reported by Simoons to the categoric protest as expressed by the Somali noble: "Speak to me not with a mouth that eats fish." "Such feelings," says Simoons, "are quite ancient in Africa, going back at least to Piankhi (c. 800 B.C.) that conqueror of Egypt who demanded that the delta princes appear before him. 'But,' Piankhi insisted, 'no fish eaters.' "[33] If we looked elsewhere in the world, we would find a host of similar avoidance patterns in almost all developing countries, affecting large proportions of their inhabitants.

Chapter 2

Early Attempts to Manufacture Fish Protein Concentrate

Although motivated by widely differing local needs, early attempts in industrial nations to produce FPC on a commercial scale encountered obstacles quite similar to those that confront contemporary FPC projects, but they do not seem to have provoked thoughtful analysis by those who planned later FPC ventures. Errors were thus repeated, often unnecessarily.

1879: GOODALE'S EXTRACT OF FISH AND ROSING'S FISH FLOUR

The following passage is found in part V of the Report of the Commissioner for 1877, issued by the U.S. Commission of Fish and Fisheries:

Goodale's Extract of Fish
The Honorary S. L. Goodale of Saco, Me. (secretary of the Maine Board of Agriculture from 1856–1873) has invented a process by which the juices of the flesh of fish are extracted to form an article of food which promises to be of much commercial value. He writes: "Some time since the idea was conceived by me and reduced to practice of concentrating the juices of the flesh of fish into a food extract. The attempts were successful . . . the product . . . bearing close resemblance to Liebig's extractum carnis[1] . . . (The) clupeidae appear to be better adapted having more nearly a simply rich-cooked meat flavor . . . No one needs less than yourself to be told how great are the possibilities for this new project. From each barrel of menhaden as taken, I get three pounds of extract when flesh alone is used and four pounds if the spine is retained in dressing. I cannot avoid the conclusion that a new source of food is within reach which at no distant day may contribute materially to human welfare.[2]

The following note is found a little further on, in the same document:

In the Norwegian Department of Agricultural Hall at the International Exhibition of 1876 were exhibited some biscuits made from "fish flour," a preparation invented by the late Anton Rosing, a prominent agricultural chemist of Norway. These biscuits were in good condition after having been kept for ten years in an unsealed jar.[3]

As far as is known, this is the first time that the term "fish flour" is used in an official document. It is also possibly the first time that fish was considered and used in the West as a suitable additive to another staple food.

1890: WAAGE'S FISH POWDER

The first attempt to produce and market fish protein concentrate on a commercial scale for human consumption in Norway and probably in Europe was made by Waage in the 1890s.[4] Professor Waage was motivated in this project by the desire to produce an inexpensive food item from fish in the form of a flour that would have both good storage properties and be of agreeable taste. He hoped to produce a material (for shipboard consumption) that would be easy to use in preparing different dishes without imparting to them the usual odor and taste of dried fish. Waage then must have been the originator of the modern FPC concept in its fullest sense. For his experiments he used coarsely chopped haddock fillets which he dried in a high-velocity stream of hot air at temperatures of 40° to 60° C until the moisture content of the processed materials reached 5 to 6 percent by weight. Although the resulting powder retained its lipid content (there was, apparently, no fat extraction stage involved in the process), it was particularly useful for the preparation of meals on board ship and was manufactured in a small plant for about four years; after this the plant was closed for reasons that are unknown.

We have no indications of attempts to produce FPC or FPC-like products on a larger scale in Norway until the 1930s when difficult conditions for the marketing of fish prompted many attempts to find new ways to dispose of surplus stocks.

1934: WIKING EIWEISS

T. Hiltner took out at least one German patent in 1934 for the manufacture of "albumen from fish";[5] many more patents were subsequently granted for this process. The idea of encouraging and supporting the development of processes for the manufacture of egg albumen substitutes was initiated by the Nazi government and culminated in the construction of at least one fish processing plant in Wesermünde. The motivation of the Germans in support of this venture was clearly to discourage importation of foreign goods and the expenditure of foreign exchange and to encourage a development that would make Germany flexible and self-sufficient in terms of her food supplies.

The fish protein product that resulted from the efforts of Hiltner and others was said to have properties equal to and, in some cases, superior to those of egg protein and could thus be offered as an egg protein substitute to all industries using that material. The new fish-based protein was also reputed to have a high nutritional value.

In 1934, the Hermann Goering Group of industries, a quasi-governmental organization, formed the Deutsche Eiweiss Gesellschaft and saw to it that the German government placed orders to assist the company while the product was becoming established. Imports of egg protein into Germany decreased rapidly as a result of these efforts while the production of fish protein, called Wiking Eiweiss, increased. In the early stages of the war, when Germany had overrun large parts of Europe, no general shortage of eggs and egg products existed for the German people. Eggs were simply collected wherever they were found and shipped back to Germany. As the war turned against Germany, however, it became essential to utilize all local protein sources, including substitutes, to compensate for the loss of food supplies. The production of egg protein substitutes reached a peak during 1943/1944. The protein products were in all cases manufactured from by-products or waste products of other industries.[6] Several thousand metric tons of Wiking Eiweiss were produced and consumed during the war years, all of which had been manufactured by the Deutsche Eiweiss Gesellschaft MBH in Wesermünde. The process was covered by twenty-five German patents, with fifteen applications pending in 1945.

The production capacity of the plant was 100 metric tons of fish protein per month. The average prewar output was 50 to 60 tons per month. The price to consumers was 9 Reichsmark per kilogram, and 6½ Reichsmark per kilogram to dealers (equivalent to \$1.64 and \$1.18/lb) respectively.

Prior to the war the Eiweiss plant used an unsalable fresh fish as raw material. During the war the raw material consisted mostly of air-dried Norwegian cod of inferior quality. Another raw material that was used during the war was shrimp, 70,000 tons of which were caught off the Dutch coast and processed. The general manufacturing process for all raw materials consisted of the following steps: treatment of the raw material with a dilute aqueous solution of acetic acid, washing and removal of water by pressing, fat extraction with 95 percent ethyl alcohol, solvent removal and recovery, pulverization of the fat-free material and solubilization of the protein with caustic soda, neutralization with acetic acid, and spray drying of the neutral material.[7]

1936: NORWEGIAN FISH POWDER CORPORATION

The Norwegian Fish Powder Corporation was founded in 1936 specifically for the production of FPC.[8] The plant built by the corporation had a throughput capacity of about 5 metric tons of raw material per 24 hours and operated mainly with haddock and cod muscle. This material was first coagulated by cooking with live steam after which the mass was processed to remove glue (collagen) and presumably some of the lipids. The remaining press cake was batch-dried for one hour in a steam-heated vacuum drier at a temperature of 40° C. In this way a fluffy, attractive product was obtained from which the major bone component was largely sifted out. The bones were ground in a disintegrator and mixed back into the fluffy powder. In spite of the fact that a considerable portion of the raw material's original lipid fraction must have remained in the finished product (no composition data are presently available) and in spite of grittiness due to the admixed bone fraction, the product was highly satisfactory in quality. Although test market investigations were encouraging, the production plant never moved beyond the trial state and was closed down within a short time.

Somewhat later, a good deal of work was done at the Norwegian Fisheries Research Institute in Bergen to find new processes to convert ever-increasing amounts of filleting waste into food products instead of using it as animal feed. One of the projects to which much attention was given was designed to produce a flour suitable for admixing to wheat flour for the manufacture of bread. Instead of cooking, pressing, and drying the press cake, which had been done previously (requiring expensive machinery and giving low yields), the raw fish muscle was finely divided and thoroughly mixed with partially dried material. This mixture was then dried without difficulty in direct-fired hot gas driers. When washed white fish muscle was used as a raw material, a light colored flour was obtained which had only a slightly fishy odor and taste. Trial production of this flour for human consumption began in 1939 and showed that a mixture of 5 to 6 percent of this material could be used in wheat flour and remain undetected in bread by the majority of its consumers. Plans to utilize the flour on a commercial scale in bread were not realized, probably due to the outbreak of war.

1937: SOUTH AFRICA

The developments in the area of FPC manufacture in South Africa were initiated in 1937 by G. M. Dreosti of the Fishing Industry Research Institute in collaboration with F. S. Mendel in an attempt to improve the protein intake of certain native population groups. Initially, a product was prepared that was virtually bland but which tended to develop a slightly fishy or bitter taste upon prolonged storage. The project was discontinued due to lack of public interest.

1938: ICELAND

Two attempts were made to produce FPC in Iceland, the earlier one in 1938. According to Dr. T. Thorbjarnarson, Director of the Icelandic Fisheries Laboratory, the raw material used for this first project was skinned cod fillets. These were vacuum dried and ground to a meal of about the coarseness of a medium-fine commercial fish meal. The process did not include a step to extract lipids and/or odor bodies. The finished product was used experimentally in fish soups, fish balls, and the baking of bread. Attempts to export this material failed.

1951: SOUTH AFRICA

In 1951, the South African Minister of Health announced plans to develop products for the protein enrichment of staple foods and two organizations announced that they could offer a product capable of meeting government specifications, that is, one sufficiently bland, when added to brown bread, to remain undetectable. One of these institutions was Dreosti's, the Fishing Industry Research Institute (FIRI); the other was the Marine Oil Refiners (MOR), both privately supported enterprises. (A product of MOR was later used for acceptability testing in Chile by the Food and Agriculture Organization of the United Nations.) Whereas the MOR organization used a pressure extraction process operating at high temperatures, the FIRI subjected the raw material to an ethyl alcohol extraction at atmospheric pressure. FIRI found that its process largely preserved the biological value of the raw material protein, while the MOR process resulted in substantial damage to the nutritional value of the raw material.

Studies conducted at the Clinical Research Unit of the University of Cape Town, in which the suitability of the FIRI FPC for the treatment of convalescent cases of kwashiorkor was tested, showed that mixtures containing 10 parts of FPC, 60 of maize, and 30 of pea flour with a total protein content of 20 percent by weight, resulted in nitrogen retentions not different from that of a pure milk diet at protein intake levels both below and above 2.5 grams per kilogram body weight per day.

These mixtures, in spite of their nutritional excellence, could not be widely introduced to the population in need of protein supplementation since a Maillard-type browning reaction changed the customary color of the cereal. This color change made the mixture unacceptable to the intended consumers and the project was consequently terminated.

1952: ICELAND

The second attempt to create an FPC operation in Iceland was made during the period from 1952 to 1954. A large fish meal plant (FAX) had been constructed in Reykjavík between 1948 and 1950; the plant was specifically designed for the treatment of herring (675 metric tons per 24 hours) and incorporated a number of modern processing techniques, particularly a dehydration/extraction step under vacuum, using a heat transfer method similar to that invented by Carver-Greenfield. Because of a succession of poor herring seasons the plant was never operated on a full-time basis. Since the plant had been designed along modern and sanitary lines, the plant engineers and chemists decided to experiment with the manufacture of FPC in the hope of eventually utilizing some of the plant equipment for this purpose. The experiments were designed to utilize a number of processing units obtained from abroad and gave fair results.

The idea of attempting to manufacture FPC was, in this case, an afterthought arising from the availability of ideal plant capacity and, presumably, from the desire to produce a material that would command a higher price and yield greater profits than animal feed. Lack of interest on the part of the fish meal manufacturers and lack of a market for FPC, however, caused the experiments to be discontinued.

PART II
THE HISTORY OF POLICY ISSUES PERTAINING TO THE GROWTH AND DEVELOPMENT OF FPC IN THE UNITED STATES

Chapter 3
FPC In the United States: Bureaucratic Politics and the Technological Imperative

The initiative to promote the development of fish protein concentrate in the United States represents an unusual but illustrative example of the difficulties inherent in the application of technological solutions to nutrition problems. While alternating between bright hope and bitter frustration, the U.S. FPC program—pursued largely by the federal government—also illustrates circumstances in which the basic inability and/or unwillingness of a governmental bureaucracy to act effectively and expeditiously on certain interjurisdictional matters (which may, in some cases, be in conflict with a particular agency's mission, perceived or mandated) caused the failure of the undertaking. In addition, the FPC case illustrates the danger of exposing a new, untested technology to the political arena before its problems and potential have been adequately assessed.

What makes the FPC story particularly fascinating, from a policy perspective, is the large number of individuals and institutions—each with its own set of priorities and motivations—that had an impact on the character and destiny of the FPC program. However, the logistics of examining this multitude of people, institutions, and relationships are not insignificant.[1] While it is important to observe the development of the program from a chronological perspective, the entry, exit, and interaction of the actors is rarely straightforward enough to permit an orderly and factual account. More often than not, events and decisions occurred in such fashion that it is sometimes necessary to draw many threads forward simultaneously in order to fully comprehend the causes and implications of any one event.

The narrative is thus presented according to the following scheme:

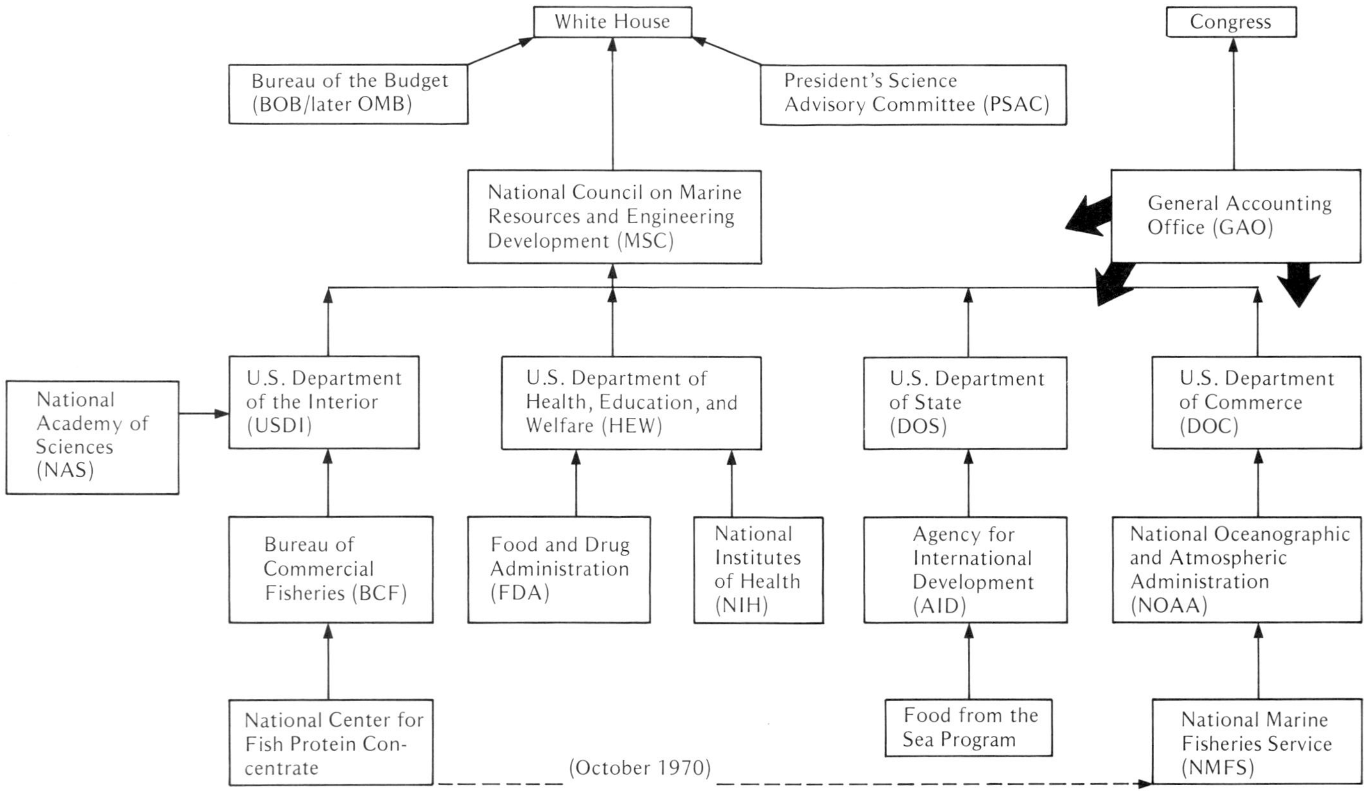

Exhibit 1
Federal Agencies Involved with U.S. FPC Program

Program Initiation (1961–1963). The sources of interest, motivation, and support for FPC in the United States during its earliest stages are considered along with the first indications of resistance to the establishment of the program. The initial public sector response, the program plan, begins to take shape.

Program Development (1963–1966). Basic policy decisions are made at this time by the promoters (public and private), the regulators (the Food and Drug Administration [FDA]), and by other actors (the Marine Sciences Council [MSC], the National Academy of Sciences [NAS], the Congress). The resolution of basic issues that emerge determines the fate of the program.

Program Implementation (1966–1969). Congress approves the construction of a demonstration plant and the testing of the product in overseas feeding programs. This stage is marked by efforts to overcome remaining obstacles, principally with the FDA, and to adequately demonstrate the potential of FPC as an effective source of low-cost protein.

Program Decay (1969–1972). Things fall apart: difficulties with the construction and operation of the demonstration plant; setbacks in the Agency for International Development (AID) program; collapse of interinstitutional support; and growing disillusionment in the Congress.

Program Termination (1972). The factors contributing to the termination of public sector involvement in the development of FPC as a protein supplement are examined, including the particular manner in which bureaucratic obstructions and program failure combined to remove political support for the program and to ensure its demise.

Epilogue (1972–1975). The tying up of loose ends, including private initiatives in the aftermath of the federal program as well as the eventual relaxation of the obstructing FDA regulations.

PROGRAM INITIATION (1961–1963): THE MOBILIZATION OF INTEREST AND SUPPORT

Public Sector Involvement and Challenge

The fish protein concentrate concept appeared on the scene at an unusual moment in U.S. history. A new President, John F. Kennedy, took office in 1961, promising a "new frontier" in domestic and foreign policy and bringing with him to Washington an extraordinary number of young, bright, and aggressive individuals. The space race, which had developed in the years following the launching of the Russian Sputnik satellite in 1957, was fully under way representing the most massive mobilization of people, technology, and resources since World War II. The enormous success of the NASA space program provided the model for emulation by many other federal agencies where staffers were often encouraged, and occasionally even expected, to adopt a similar "blue-sky" approach regarding program formulation. The most significant characteristic of the period, however, was that the kinds of constraints on capital, energy, raw materials, and other resources which have become so common in the 1970s were virtually unknown. There was, in sum, a pervasive feeling that all problems were solvable simply through the appropriate application of human and fiscal resources, although now, with the benefit of historical perspective, it is apparent that this approach was far better suited to pure technological applications such as the lunar landing program than to people-oriented problems such as malnutrition or poverty.

One of the youngest members of the new Kennedy cabinet was the Secretary of the Interior, Stewart Udall. Udall's departmental policy, conforming to the Kennedy "new frontier" theme, strongly urged that new programs be sought through which the department could become involved in the spirit of the new administration.[2] As a result, the Bureau of Commercial Fisheries (BCF), a line agency of the U.S. Department of the Interior (USDI), seized upon the recent endorsement of FPC by the Food and Agriculture Organization (FAO) of the United Nations[3] to promote U.S. involvement in the development of the product.

The BCF had been created out of the Fish and Wildlife Service in 1956. After the commercial and sports fisheries agencies were separated, an intensive effort was begun, lasting into the Kennedy administration, to examine the proper role for the commercial fisheries agency. FPC was viewed as one program through which to focus the agency's mission.[4] The Bureau believed that in the long run FPC might (1) encourage increased utilization and upgrading of ocean resources; (2) potentially improve the economic status of the U.S. fishing industry; and (3) enhance the world image of the United States by helping to alleviate worldwide protein malnutrition.[5] A more immediate concern on the part of the Bureau related to the depressed state of the U.S. fish meal industry. It was anticipated that the development of FPC might facilitate the transformation of slack industrial capacity to the production of a human-grade product.

The possibility of USDI involvement with a project such as FPC quickly caught the interest and imagination of several key individuals on Secretary Udall's staff, including James Carr, then Undersecretary of the Interior, and Dr. Roger Revelle, the newly appointed science advisor to the Department. The role of scientific advice in the early USDI decisions regarding an FPC development program was, in fact, substantial since not only did Revelle have direct, informal access to Secretary Udall, but Dr. Jerome Weisner, Special Assistant to the President for Science and Technology and Chairman of the President's Science Advisory Committee, was also active in promoting the FPC concept through contacts in the Congress and the federal bureaucracy.

Undersecretary Carr became the principal link between the preliminary development efforts at the BCF and the top USDI management.[6]

As a result of this awakening interest in the USDI, the fiscal 1962 budget message presented to the Congress reflected the Kennedy Administration's newly developed awareness of the potential of the ocean's resources. Mainly representatives from states with fishing industries, Congressional advocates of FPC were able to authorize an appropriation for FY 1962 of $50,000, under authority of the Fish and Wildlife Act of 1956. The program initiated by the BCF was to consist of three phases: (1) a comprehensive literature search and the establishment of an FPC data base; (2) a worldwide survey of all known and existing processes for the manufacture of FPC; and (3) a laboratory research program to develop commercially useful methods for FPC production.

The largest and most expensive phase of the program would be the laboratory development of a satisfactory process, either by improving an existing method or by developing a new one. BCF researchers expected that any such process developed would have many or all of the following characteristics:

Require low capitalization

Be economical in operation

Have flexibility to permit large- or small-scale operation

Have flexibility to permit operation in parts of the world where utilities were limited

Produce an end product that was acceptable in varying cultures

Produce an end product that was approved by nutritionists

Produce an end product that could be cheaply shipped, stored for varying amounts of time, and easily incorporated into local diets

A basic principle of the plan was that it was not to place the federal government in competition with private industry in the production and marketing of any product which was developed. Quite the contrary, it was anticipated that the BCF would merely lend technical and financial assistance until a process existed which industry could take over. The Bureau's research program commenced in July 1961 at its College Park, Maryland, research facility. These efforts were designed to investigate simultaneously three separate modes of production: chemical, biological, and physical.[7] Preliminary results from the literature search and process survey indicated that ample justification existed for proceeding with a basic research and development (R&D) program either to suggest necessary modifications of the few processes already extant or to create an entirely new process. Two particular extraction methods had received significant attention by Bureau scientists: the single-solvent process developed in Canada and the dual-solvent process invented by the VioBin Corporation of Monticello, Illinois.[8] After examining the various methods from a variety of perspectives, such as nutrition, industrial viability, solvent recovery, and toxicity, the consensus of the College Park researchers was that the Canadian single-solvent process offered a safer, more reliable technique.

Opting for the Canadian process was strongly disputed by Ezra Levin, president and founder of the VioBin Corporation, who argued that his dual-solvent process possessed all of the necessary characteristics, was safe and reliable, and, most importantly, was already available. Levin saw little justification on this basis for a federal agency to pursue publicly supported research on a product that his company had already developed with private capital.[9] Other factors that entered into the BCF decision to bypass the VioBin process included (1) the concerns about the safety and reliability of the VioBin process; (2) the prior history of conflict between the VioBin Corporation and the Food and Drug Administration; and (3) increasing political pressure on the USDI to create and implement their own new ideas for the Kennedy "new frontier" image.

The VioBin Corporation had undertaken informal discussions with the FDA as early as 1959 in order to establish a standard of identity for FPC as a food product. But, when the regulatory agency indicated, informally, that it was likely to find the product adulterated, Levin approached certain members of Congress, namely, Senators Paul Douglas (Illinois) and Leverett Saltonstall (Massachusetts). Both Senators were inclined to be sympathetic to Levin's problem, the former representing the home state of the VioBin Corporation (and a personal friend of Levin), and the latter representing one of the principal eastern fishing states. The National Fish Meal and Oil Association (NFMOA), a fishing industry trade association, had also become concerned about the position of the FDA regarding the acceptability of FPC processed from whole fish. When an aide to Senator Saltonstall filed a petition with the regulatory body in July 1961 for a standard of identity for whole FPC on behalf of the VioBin Corporation, the issue was joined.

FDA Response: The "Filth" Issue

In responding to the request, the FDA expressed the informal opinion that the VioBin product violated Section 402 (1) Clause 3[10] of the Food, Drug and Cosmetic Act of 1934 in that it contained viscera, bones, scale, and so on, which were regarded as filth, and thus the product would be considered adulterated and would be prohibited from sale in the United States. This opinion also stated that, under Section 801 (d) of the Food, Drug and Cosmetic Act, the product could still be sold outside the United States, where relevant national laws permitted the sale.[11] Such an occurrence did not seem likely, however, in light of the domestic prohibition since it would be potentially disastrous for a government to approve for consumption a food that had been banned in the United States specifically on the basis of its impure contents.

In response to industry inquiries about the meaning and likely ramifications of its "informal" opinion, FDA officials suggested three possible courses of action which might be adopted with regard to FPC:

1. Industry could seek redress from the Congress in the form of an amendment to Section 402 (a)(3) of the Food, Drug and Cosmetic Act specifically permitting the use of whole fish.
2. The product could be marketed and shipped in interstate commerce in the United States; whereupon it would be seized as an adulterated food by the FDA, and subsequently tested in court.
3. The FDA could be petitioned to establish a definition and standard of identity that would regulate the product.

The petitioners considered the first alternative[12] extremely unlikely, especially at such an early stage in the development of the product, while the second involved tremendous uncertainty, costly delays, and much negative publicity. Thus, industry representatives and the BCF, which had been drawn into the matter due to the VioBin Corporation's preemptory action, determined to pursue the standard of identity. The VioBin request, together with the informal opinion by the FDA, was duly published in the *Federal Register* on September 15, 1961. Pursuant to the regulations of Section 401 of the Food, Drug and Cosmetic Act, the petition sought to establish a Definition and Standard of Identity for a product called "fish protein concentrate, whole fish flour."[13]

During the sixty-day period required for comments on a proposed new product, 1870 responses to the petition were received by the FDA. These originated primarily from the fact sheet on FPC that the agency had made available to its Consumer Consultant Network to solicit opinions. Although the majority of the reactions were favorable to the whole-fish product (1038 for and 832 against), the opposing comments were significant enough to merit further FDA consideration. Predictably, many of the objections cited the fact that the product would include portions of the fish not normally used as food. Baking, milling, and dairy groups were opposed to FPC on the grounds that it might cast doubt on the purity of their own products, which were prepared under tightly controlled product component and sanitary conditions. Protests were also received from Congressmen of some of the major agricultural states (Kansas, Wisconsin, North and South Dakota) as a result of constituent pressure from dairy and wheat farmers. The departments of health and agriculture in most of these states also voiced their opposition. The AFL-CIO, representing the interests of organized labor, wrote that this was an atrocious proposition that would add bulk to a loaf of bread and raise its price to the working man.[14]

Through a long series of judicial and administrative decisions, the FDA had worked to create a precedent which established that the incorporation in a product of ingredients generally considered filthy would make the product unacceptable, even if the product itself were judged safe and nontoxic. (For elaboration of this issue, see chapter 5.) The regulatory agency thus viewed the approval of FPC—solely on the basis of its safety and nutritional content and without regard to purity or aesthetic acceptability—as counterprecedent-setting in that it would open the door to the reconsideration of other products not approved because of their unsuitable composition. Thus, although no scientific research on the product had been conducted or requested by the agency, on January 25, 1962, FDA Commissioner George Larrick, citing both the formal objections to the 1961 proposal and Section 402 (a) Clause 3 of the Food, Drug and Cosmetic Act, issued a modified definition and standard of identity that did not permit the use of whole fish.

Shortly after the publication of the FDA's proposed standard of identity for FPC, formal objections were filed by a variety of industrial concerns and trade groups: the VioBin Corporation (Illinois), the Gulf Menhaden Company (Louisiana), the Industrial Products Division of the National Fisheries Institute, the Virginia Fisherman's Association, and the Fish Products Company (Delaware). Senator Douglas (Illinois) also filed an objection. Attorneys representing the VioBin Corporation made an additional protest to what they considered improper activities on the part of the FDA in the form of press releases, statements, and interviews which were designed to prejudice the position of various consumer and health groups on the issue.[15] As a result of these objections, the FDA stayed its order on FPC and scheduled hearings on the matter for June 18, 1962.

National Academy of Sciences Study

In the meantime, four new bills had been introduced in the House of Representatives to amend Clause 3 of Section 402 (a) to permit FPC to be made from whole fish, provided that the fish were safe and handled in a sanitary manner.[16] In addition, a committee of the NFMOA, after careful study, had decided that the known processes for producing FPC were not yet well enough perfected to undergo FDA scrutiny. The committee thus agreed to refer the matter to Secretary of the Interior Udall in the hope that he would invite the National Academy of Sciences-National Research Council (NAS) to initiate a study of the issue.

The Interior Secretary responded by requesting that an ad hoc committee of the NAS be appointed to review the safety of whole fish protein concentrate and to make recommendations for appropriate action.[17] It was anticipated by the Interior Department that the support of the prestigious National Academy of Sciences would be helpful in attracting greater attention to the program, which might result in a larger congressional appropriation. In addition, a favorable NAS report might help to provide leverage in the struggle to win FDA approval of whole FPC.[18]

Secretary Udall requested that the ad hoc committee pursue the answers to three specific questions: (1) Is FPC wholesome, safe, and nutritious? (2) Does a suitable product already exist? (3) Is there a demonstrable economic and nutritional need for such an inexpensive animal food supplement in the United States? It is curious that Udall's last question was framed in such a way as to emphasize need in the United States rather than the developing world. Perhaps the only explanation is the fact that the product required approval by the U.S. Food and Drug Administration, and thus proof of need in the United States became significant. The concept of need was confused with demand here, as was often the case in the course of FPC development.

Attorneys for the NFMOA used a copy of Udall's mandate to the NAS and the existence of the bills in Congress as evidence in support of their request to the FDA for an indefinite postponement of the June 18 hearings. The FDA responded by formally accepting the request and ordering the hearings stayed indefinitely pending the outcome of the NAS review.[19]

Udall himself had made an attempt to bring the parties together and to emphasize to the FDA his involvement in and commitment to the program. On May 10, 1962, the Interior Secretary had invited Health, Education and Welfare Secretary Ribicoff, FDA Commissioner Larrick, presidential science advisor Wiesner, certain Senators and Congressmen, and other select individuals to a private luncheon in his office. Among those in attendance was Eunice Kennedy Shriver, the President's sister and the founder of the Eunice Kennedy Shriver Center for Mental Retardation, Inc., who saw in FPC a great potential for preventing the mental retardation caused by malnutrition in children of the developing world. Udall, Shriver, and the members of Congress joined in pressing Ribicoff and Larrick for the reasons for FDA resistance. Secretary Ribicoff responded by restating the substantive objections raised previously by the regulatory agency. However, when it was Commissioner Larrick's turn to respond, he produced a series of glossy color photographs depicting in great detail the sight of a fish with its gut cut open and its viscera protruding. His response was: "Do you want people to eat this?"[20]

Despite Larrick's continuing crusade against "filthy fish," the FDA did act to stay the hearings. This led the Department of Interior, in an effort to avoid further antagonizing the FDA, to urge that the four bills then under consideration by the House[21] also be tabled until the NAS committee report was completed. The House Subcommittee was willing to comply with the USDI's request but by this time thirty-two witnesses had already presented testimony; fifteen were in favor of FPC, citing the worldwide need, the nutritional advantages, the availability of cheap fish, and the potential to boost fishing (and fishing-related) industries; and seventeen were opposed, citing the questionable need for FPC in the United States, the importance of maintaining aesthetic and purity standards, and the danger of weakening the regulatory power of the FDA.[22]

Under the joint auspices of the Division of Biology and Agriculture and the Food and Nutrition Board of the National Academy of Sciences, an ad hoc committee was established on the subject of FPC and work commenced during the summer of 1962. By the end of October, the ad hoc NAS committee was prepared to make public its response to the questions proposed by Secretary Udall. The group found that, while an acceptable commercial method of FPC production had not yet been developed, a safe, wholesome, and nutritious product could, in fact, be made. The committee also concluded that no demonstrable need for a new animal protein food additive then existed in the United States, but that this fact should not preclude further research on new foodstuffs. Thus, its primary recommendation was that steps should be taken by the BCF to develop a viable commercial process; and on that basis it suggested that FDA hearings be stayed indefinitely until a suitable process was developed and the product was adequately tested. The NAS committee also cited the need for a permanent scientific advisory group to aid and advise the BCF on marine protein research.

Pursuant to the recommendation of the ad hoc committee, the National Academy of Sciences, under contract to the Department of the Interior, did move to create a permanent committee of the Food and Nutrition Board in early 1963. Known as the Scientific Advisory Committee on Marine Protein Resource Development, this group was to play an active role not only in advising the BCF on its continuing R&D efforts but also in acting as an advocate (and legitimizing influence) for FPC in its consideration by the FDA.[23]

PROGRAM DEVELOPMENT (1963–1966): THE EMERGENCE OF THE ACTORS AND THE ISSUES

Public Sector: Program Acceleration

Partially in response to the principal recommendation of the NAS Committee, the Kennedy Administration requested an FY '64 $400,000 supplemental appropriation for the Bureau of Commercial Fisheries to facilitate scaled-up development of processing methods to manufacture a high-quality FPC. In April 1963, Congress approved the supplemental funds, thereby raising the total FPC budget to $520,000, a tenfold increase in the course of a single year. With only two months notice before the beginning of the expanded FPC funding, officials at the BCF scrambled to negotiate contracts with various engineering organizations and universities. In June, contracts were set for the design and construction of the pilot scale equipment for both the chemical (solvent extraction) and biological (enzymatic hydrolysis) processes.

At the start of the new fiscal year in July 1963 Interior Secretary Udall, recognizing the growing urgency of the world protein problem relative to the FDA position, ordered a major policy shift in the BCF research program. He directed that development be accelerated by concentrating on one extraction method only. The method finally chosen was to be exhaustively tested and then presented for approval by the FDA. This decision was reached basically for pragmatic political reasons. Unlike another "high technology" USDI program (the effort to perfect a cost-effective method of water desalinization for which Congress had funded five different demonstration plants in various parts of the country), this FPC program did not elicit a convincing degree of enthusiasm and commitment from lawmakers or from the general public. For this reason Udall saw a fundamental need to produce a viable product at the earliest possible time in order to consolidate and strengthen congressional support. The policy shift was designed, therefore, to speed up program development by focusing available funds more narrowly.[24]

Private Sector Initiatives

These developments at the BCF also had their parallel within the private sector. Besides the continuing efforts of the VioBin Corporation to gain FDA approval for its own solvent extraction process, there were initiatives by other major industrial concerns to commit significant amounts of private R&D funds to the search for a commercially viable FPC process. General Foods Corporation developed a chemical extraction process that it patented in Canada in May 1963, and later in the same year, General Foods received a patent in the United States covering the isolation of a heat-coagulated protein from fresh fish by means of an aqueous extraction process.[25] The J. Howard Smith Corporation, a leading processer of menhaden for fish meal, was also active in experimental FPC research during this period.

It seems fair to question whether the initial decision by the BCF to pursue the bulk of FPC research primarily through its own in-house laboratory at College Park, Maryland, rather than through private contract, was ultimately a wise one. In the competitive world of the private sector, companies that manufacture a product of poor quality or suffer large cost overruns must, with certain notable exceptions (for example, the decision by Congress to assist the Lockheed Corporation in its financial difficulties relating to the C-5A contract), pay for their poor performance through the market mechanism. Government laboratories, on the other hand, have a tendency to continue in operation despite the occasionally poor quality of the work produced. It may be argued in the case of FPC that, despite the R&D initiative cited, the necessary interest and/or expertise did not exist at that time within the private sector (with the notable exception of the product marketed by the Vio-Bin Corporation) and that development within the government laboratory was the only reasonable policy course available. Yet, the fact remains that development of FPC through competitive bidding was apparently never considered seriously within the BCF and, therefore, the likely outcome of this policy option remains unknown.

BCF R&D Policy Decision: Process and Raw Material

Most of the promising FPC process developments as of 1963 had utilized some form of chemical extraction. The Canadian process, developed at the Halifax Laboratory of the Fisheries Research Board of Canada,[26] showed particular promise using isopropyl alcohol (IPA). It was on the basis of the recent Canadian experience that the IPA solvent extraction process was chosen by the BCF as the most reliable and potentially expedient method of producing FPC. Although a modest amount of money remained committed to research on biological (enzymatic) and physical processes, the bulk of the funds available was shifted into a crash effort focused on solvent extraction.

Another major decision facing the researchers at the BCF's College Park laboratory concerned the type of fish to be used as the principal raw material. Because of the additional engineering problems involved in the extraction of the so-called oily or fatty fishes, a decision was made to concentrate attention on lean fish. Accordingly, the red hake, an underutilized lean fish found in abundance off the northeastern coast of the United States, was selected as the initial target. The hake was selected primarily because (1) it was a schooling fish and thus more likely to be accepted by the FDA than a mixed species;[27] (2) it was underutilized; and (3) it was easily available to the College Park laboratory from east coast ports.

Both of these BCF decisions—the choice of the IPA solvent extraction method and the selection of red hake—were to become important determinants in subsequent decisions made by the Bureau. Debate over the choice of the solvent extraction process began almost immediately, due to the insistence of Ezra Levin that his dual-solvent process was superior. Later, FDA concerns about IPA residues (resulting from the use of the solvent) eventually required the addition of expensive desolventizers and other equipment that significantly increased the cost of the finished product.[28]

The choice of red hake, a lean fish, has been a matter of controversy primarily because, as the Bureau was aware even in 1963, the greatest potential resource for FPC manufacture are the so-called fatty fishes, which constitute the majority of the underutilized (and, therefore, cheap and abundant) species in the world. While it was true that lean fish presented far fewer technological complications than fatty species, the decision *not* to deal initially with those complications led subsequently to serious miscalculations in both cost and design of larger-scale processing equipment and the time necessary to deal with the problems.

Senate Hearings on FPC Viability (1964)

On August 14, 1964, the Congress took up the FPC question directly in hearings before the Subcommittee on Merchant Marine and Fisheries of the Senate Committee on Commerce, chaired by Senator E. L. Bartlett of Alaska. The hearings were held, according to Bartlett, "to learn what standards a marketable concentrate would have to meet, how the products the Bureau has developed stand in relation to those standards, and how expensive or difficult they would be to produce."[29]

In his prepared statement for the Subcommittee BCF Director Donald McKernan indicated that FPC development in the United States might meet the following objectives: (1) improvement of the U.S. world image in the fight against malnutrition; (2) usefulness in the domestic antipoverty campaign; (3) conservation of fishery resources; (4) support for domestic fishing industries through diversification, and (5) contribution to other related industries such as shipbuilding and industrial processing.[30] McKernan reported to the Senators that the FDA still opposed the use of whole fish as a violation of the Federal Food, Drug and Cosmetic Act, but that the differences between the positions of the two agencies could probably be resolved. Also at the hearing, Donald Snyder, Director of the College Park Laboratory, predicted, in response to a question by Senator Bartlett, that from the time the model scale unit (then under construction near College Park) became operable, it would take only one year until at least one of the processes would be commercially viable.[31] Subsequent events would prove this prognostication extremely optimistic.

It is clear that, even as late as the 1964 hearings, the FPC concept was being "oversold." There were still no firm cost or time estimations available, and the constant shuffling of figures by various BCF officials was a cause of great consternation both in Congress and the Bureau of the Budget.[32] Because the program had been promoted on the basis of incomplete information, the government laboratory was forced to constantly scramble to provide technical responses to questions on aspects of the program that were often a number of steps ahead of the current state of research. The failure of certain of these "guesstimations" proved extremely embarrassing and injurious to the program. The project was also hurt by the departure in late 1964 of two staunch FPC advocates within the USDI, James Carr and Roger Revelle. In their absence the approach adopted toward FPC within USDI became increasingly more political, that is, more public-relations oriented, and less technical.[33]

NAS Recommendations: FPC Quality and Production

On September 16 and 17, 1965, the NAS Committee on Marine Protein Resource Development met for the third time to review the results of the College Park Laboratory's work with red hake.[34] Using the data submitted by the BCF, the NAS committee was able to state categorically that, in its opinion, "the FPC developed by the Bureau of Commercial Fisheries is wholesome, highly nutritious, stable, safe and capable of economic production and marketing."[35] Further, in a strongly worded response to the FDA filth objection, the committee stated:

> In the judgment of the committee, FPC, as produced by the BCF process, is not in any way filthy or otherwise unfit for human consumption. On the contrary, after reviewing in detail the BCF report on the hake project . . . the Committee has concluded that the FPC made by the BCF process is a product that is wholesome and safe for human consumption, is highly nutritious, blends well as an additive to many types of food normally used, can be produced at a reasonably low cost and, in the view of the Committee, is suitable at present for commercial production, distribution, and use in human nutrition.[36]

These conclusions were formally adopted by the committee on October 14, 1965, at which time they also recommended the construction of an experimental pilot plant for production of FPC, suitable for human consumption, under the operational control of the BCF. It is instructive to note the careful working of the committee's statement, which indicates that it was based on no new, original research. Later, the FDA would take legitimate exception to the attempt of the BCF to represent these findings as scientific evidence on which the regulatory agency should base its rulings.

FDA Response: Policy Shift

Pressure on the FDA to revise, or at least modify, its "anti-filth" position had continued to come from many sources. VioBin Corporation President Ezra Levin maintained a steady stream of correspondence with the regulatory body. There was also an awakening interest concerning the FDA impasse on the part of Congress.[37] However, despite the availability of significant new data amassed by the BCF in the course of its research with red hake, the FDA position had remained unchanged. Pressure for resolution of the impasse had begun to mount, however, with the formal presentation of the NAS committee's report on FPC to Secretary Udall on December 1, 1965.

The FDA emphasized that the ultimate decision on the acceptability of FPC was the responsibility of the FDA only and that the endorsement of the National Academy of Sciences, while given much weight, would neither prejudice nor accelerate its decision. The FDA also revealed a major shift in policy by recommending that FPC be presented for approval as a food additive (rather than as a whole food) under the Food Additive Amendment, Section 201 (d) of the Food, Drug and Cosmetic Act of 1938. (For further details on the issues surrounding this shift, see chapter 4.)[38]

BCF Decision: The Quest for FDA Approval

The BCF leadership was thus faced with a major decision. Despite the ongoing research at College Park it was obvious that commercially significant work could not be implemented without FDA approval. But the shift from a food standard of identity to a food additive meant further delays while research efforts were reoriented. Did the potential of the program merit the additional time and money that such research would require? In search of an answer to this question (and in search of elusive financial support from industry) BCF Director McKernan called a meeting of industry representatives in January 1966. Gathered at the meeting were individuals possessing the very best available knowledge about fisheries and the realistic potential of the oceans and the fishing industry. The question put to the group was simply stated: should the federal government proceed with its FPC program, or, in other words, did it have strong potential to succeed?

The group was united in its support for the continuation of the BCF program. However, none of the industry people present was willing to be specific as to dollar amounts or timetables for anticipated private sector involvement. Various reasons were given for the reluctance of private industry to take over process development:[39] (1) continued uncertainty about the likelihood of FDA approval; (2) possible advantages of the biological process over the BCF chemical process; (3) high capital cost of the processing equipment; (4) need for further research concerning acceptable and available species of raw materials; (5) lack of information concerning the cost of the finished product, economics of scale, and so on; and (6) unanswered questions about consumer acceptance and demand.

Encouraged by even the guarded support of the industry group, the Bureau determined to pursue the FPC program with the knowledge that private industry was not likely to assume the costs or the risks of development until its substantive questions had been answered by further federally funded research. Accordingly, advice was sought from Vincent Kleinfeld, a well-known Washington food and drug attorney, regarding the pros and cons of pursuing a food additive petition as proposed by the FDA. He expressed the opinion that if FPC was wholesome and safe, it should *not* be necessary to submit a petition as a food additive since such a shift in policy might cast doubt on the status of the product as a pure food.

While the leadership of the Bureau considered the wisdom of Kleinfeld's recommendation not to pursue the food additive petition, other factors were combining to reduce the options available. On February 10, 1966, President Lyndon B. Johnson, who viewed the FPC program as yet another component of the "great society" theme, delivered his so-called "War on Hunger" message:

> The wonders of modern science must also be directed to the fight against malnutrition. I have today directed the President's Science Advisory Committee to work with the very best talent in this nation to search out new ways to develop inexpensive, high-quality synthetic foods as dietary supplements. A promising start has already been made in isolating protein sources from fish, which are in plentiful supply throughout the world.[40]

Shortly after this message was given, the long delayed report of the NAS Committee on FPC was also released to the public by the USDI.[41]

With FPC now under both presidential and public scrutiny and with nothing more than laboratory samples available for analysis and testing, it was apparent that the likelihood of receiving FDA approval for a food standard for FPC was quite remote. Thus, based on FDA assurances that a food additive petition, unlike the food standard of identity, might avoid the filth issue (since the manufacturing quality of the contents of food additives are closely regulated and labeled), the BCF opted to redirect its research toward gaining approval of FPC as a food additive. A food additive petition was duly filed on February 25, 1966, proposing FPC extraction using the IPA method.[42]

Was the BCF railroaded into pursuing the food additive petition rather than the standard of identity? Many food technologists and nutritionists had argued from the beginning that FPC, as derivative of fish, was a *food* and must be considered as such or its claim to wholesomeness would be impugned. This led a number of food and drug attorneys to suggest that the FDA policy shift was based on dubious precedent and that it should be immediately challenged in court.[43] But the Bureau was caught between the proverbial "immovable object" (the FDA) and an "irresistible force" (White House demands for action). It is important, in this regard, to reemphasize the irresistable nature of demands for action made from the Johnson White House. To resist the presidential imperative, especially after Johnson had publicly committed himself, would have been political suicide for the BCF.

In addition, the BCF suffered from the lack of ability to apply potent counterpressure both from within and outside of the federal bureaucracy. (This is the kind of pressure so effectively applied by the Department of Agriculture in its dealings with the FDA.) Since the USDA represents the U.S. farmer, it has at its command many powerful weapons (Congress, public opinion pressure, organized farm lobby groups) with which to influence the FDA's deliberations. The USDI, however, does not have the same powerful lobby support, and interest in the development of coastal fisheries was certainly not in the mainstream of its activities. Moreover, since the fishing industry itself was more concerned with the real (or imaginary) incursions of Russian and

Japanese fishing fleets into U.S. territorial waters than with new marine technology, there was no effective means of leverage against the FDA's position.[44]

Pressure for Results

After the 1963 decision by Secretary Udall directing the BCF to concentrate development efforts on the chemical extraction process, work had progressed at the bench scale in College Park, Maryland. These efforts led to the construction of the so-called Model Scale Unit, a "pilot plant" scale apparatus, in Beltsville, Maryland, which supplied the FPC (in 150-pound batches) used in most of the tests conducted to provide information required for the food additive petition. Plans were then developed for the construction of a larger Expanded Model Scale Unit, which was never built due to increasing time pressures and cost constraints. A decision was made instead to enlarge the capacity of the existing apparatus and thereby shift the production procedure from batch processing to continuous operation at a capacity of 300 pounds per hour. A private engineering consultant, Artisan Industries, Inc., of Waltham, Massachusetts, was engaged to design the specifications for the new unit, which was to be called the Modified Model Scale Unit (MMSU).

The MMSU was intended to provide the information needed for the precise design of a continuous extraction plant at the commercial scale. Many of the design requirements for continuous extraction, the only mode of manufacturing FPC that was considered economically realistic, were not yet available in 1966. However, as a direct result of the increasing pressure to demonstrate the commercial viability of FPC technology, the BCF decided to forego the MMSU entirely[45] and jump directly to the development of the final stage of the R&D process, namely, the Experiment and Demonstration Plant (EDP), a continuous extraction apparatus designed for operation at a semicommercial scale. Thus, a vital step in the development scheme was skipped and the necessary data on the requirements for continuous processing were not obtained. This kind of omission of a development phase is not that unusual for a technology pursued under public and political scrutiny,[46] when the need for success dictates a gamble, but its occurrence was to have major technical consequences in subsequent crucial stages of the BCF program.

Senate Action on the FPC Bill (1966)

The mounting political pressure for results in the FPC program became evident when the fisheries advocates on the Senate Commerce Committee, Senators Bartlett and Magnuson, cognizant of the need for testing at the industrial scale, and perhaps somewhat impatient with the pace of the BCF development program and its continuing problem with the FDA, jointly introduced S. 2720 during the second session of the 89th Congress. The "FPC bill"[47] called for the construction of five plants at various locations around the country and was modeled closely on another recent USDI program involving water desalinization.[48]

The Interior Department submitted a report favoring the measure, although it advocated the construction of only one plant instead of five. The Bureau of Commercial Fisheries, represented by Director McKernan, testified that a commercial scale, continuous extraction plant was considered essential (1) to test and confirm engineering concepts; (2) to demonstrate to U.S. fishing and food processing industries the practicality and versatility of the product; (3) to provide sufficient quantities of FPC for large-scale feeding and storage tests; (4) to conduct feasibility studies for the creation and development of markets for FPC-supplemented foods; and (5) to help determine the commercial costs of using fish in this manner.[49] The Agency for International Development also testified in favor of the measure in support of its interest in the product for overseas feeding programs.

Among the many witnesses to appear before the Commerce Committee was Congressman Hastings Keith of the fishing port of New Bedford, Massachusetts, who strongly condemned the continued FDA refusal to approve whole FPC: "I hope and believe that the facts will . . . demonstrate that the FDA has a long history of bureaucratic foot dragging and negativism in this matter." Mr. Keith compared the FDA's handling of the FPC issue with the recent "cranberry scare" in New England, whereby FDA pronouncements had caused tremendous financial loss to that industry.[50] But criticism was also expressed concerning the direction of the FPC program by the Bureau. In a letter to the Commerce Committee VioBin's Ezra Levin claimed that the BCF was "not equipped by training, experience or administrative judgment to direct such a development" and that the Committee should act to remove the direction of the FPC research from the control of the Bureau and should turn it over to the universities under federal contract.[51]

Despite Levin's criticism, and the recommendations for a reduction in the number of plants, the Commerce Committee report concluded that the development of the FPC program was justified by the population explosion and by the benefits to be derived by developing countries from a new protein source of this nature. The report, issued June 23, 1966, recommended the construction of five FPC plants in order to reduce the degree of private risk, to stimulate development of an FPC market, and to provide adequate quantities for FPC for research and testing.[52] Similar to the 1964 hearings, Congressional enthusiasm for the program, generated largely by the Bureau's own promotional efforts, restrained the BCF from revealing the actual status of research and development. Serious technical obstacles remained that, under other circumstances, might well have caused the delay, if not the termination, of plans for the demonstration plant. But too many political bridges had already been crossed to permit the Bureau to request more research time and funding. With a substantial number of jobs and professional reputations on the line, management felt obliged to plunge ahead with the hope that the unanswered questions could be resolved as the work progressed.

It is worth noting that the 1966 Senate hearings marked only one of several occasions when BCF witnesses were less than candid as to the actual state of R&D in the FPC program. It was obviously not in the best interests of the program, from the advocate's perspective, to be totally frank as to the nature and degree of the problems that had been encountered. Yet, by the same token, the relevant congressional committees must also be faulted for not doing their staff homework and for not pressing harder for a realistic assessment of the program's actual status. In the final analysis, many of Levin's criticisms of the BCF management were at least partially justified by the administrative failures and judgmental errors made by the Bureau in the course of the FPC program.

Marine Sciences Council: Formation and Mandate

Just prior to the release of the Senate Committee's report the Congress passed the Marine Resources and Engineering Development Act of 1966. This act served a twofold purpose. First, it established a Cabinet-level committee within the executive branch which was to be chaired by Vice President Hubert H. Humphrey and which included as members the secretaries of State, Navy, Interior, Commerce, HEW, and Transportation, as well as the chairman of the (now defunct) Atomic Energy Commission and the director of the National Science Foundation. The major function of the National Council on Marine Resources and Engineering Development, or the Marine Sciences Council (MSC) as it was commonly called, was to advise and make recommendations to the President on U.S. policy and budgets concerning the utilization of marine resources.[53]

The Act also created the Commission on Marine Sciences, Engineering, and Resources, composed of nineteen respected citizens from outside the government, whose purpose was to conduct a thorough analysis of the marine resource needs and potential of the United States. There was, of course, some overlap between the two bodies, but the primary mandate of each differed substantially. While the commission was created, amidst some controversy, as an ad hoc group of limited duration, the MSC had no specific statutory life. Whereas the commission was mandated to carry out a particular task, the MSC was viewed as an instrument of the President to aid in the formulation and coordination of marine resource and science policy.

By the summer of 1966, the MSC had been fully constituted and was faced with an explicit directive from President Johnson to prepare a list of specific proposals relating to the use of marine resources by January 1967. An initial survey of priority needs was conducted under the auspices of outside consultants and the various federal agencies involved in marine affairs, including, of course, the BCF. The development of FPC was selected as one of nine priority projects proposed to the President, apparently on the basis that "it could be attuned to the President's 'War on Hunger' and, therefore, it was not a matter of selling him a new program but simply a new technique."[54] Two task groups, which included representatives from both AID and the BCF, were formed in October 1966 to examine the various projects that had been proposed to the President. Since the BCF was interested in FPC for its potential to help the fishing industry and since AID already had responsibility for the War on Hunger, it is hardly surprising that the report of the task group was highly favorable to the promotion of the FPC program.[55]

House Hearings on the FPC Bill (1966)

Now it was the House of Representatives that inquired into the advisability of funding an FPC demonstration program. Hearings before the Subcommittee on Fisheries and Wildlife Conservation of the House Committee on Merchant Marine and Fisheries were convened on August 11, 1966, to begin consideration of some fifteen variations of the Senate-passed bill, S. 2720. The principal difference between the various bills related to the specific number of FPC plants recommended for authorization by the Congress.

Representative Hastings Keith (Massachusetts) wasted little time in renewing the attacks on the FDA that he had initiated in the Senate hearings: "The FDA has remained adamant in their position. After all, we do eat sardines, and quahogs and cherrystones, and some of the Japanese eat butterflies and things of that sort. We do eat lots of filth during the course of our lifetime . . . In my view, they [the FDA] are sometimes too zealous . . . "[56] Keith cited the recurring problems over the fluoride issue in relation to the previous national debate over fluoridating water supplies. He also noted the fears about FPC raised by the soy and flour industries, stemming from the early 1960s when FPC had unwisely been called "fish flour." Keith suggested that through a gradual education process the flour industry had come to view FPC as an acceptable wheat fortificant, rather than as a competitor.

BCF Director Donald McKernan was among the first witnesses appearing before the subcommittee. He repeated many of the same argumants presented to the Senate Commerce Committee, along with the additional justification that the FPC plants would extend the fishing season and thereby raise the income of fishermen. McKernan reiterated his support for the funding of a single pilot plant, but he cautioned that this would place limitations on effective testing because of the need for additional equipment for processing fatty fish.[57] He indicated that the BCF would expect to produce 3000 tons of FPC per year for about two to three years.[58] During the House hearings, as had been the case with the Senate investigations, little or no mention was made by the BCF witnesses of the various problems that had been encountered in the R&D process.

Subcommittee Chairman John Dingell (Michigan) expressed his reservations to McKernan about approving the construction of a pilot plant when the possibility existed that the FDA might reject the BCF food additive petition. This concern was repeated by the leading conservative member of the committee, Alton Lennon (North Carolina), who also questioned the propriety of using public funds for the development of a process later to be used for profit by private industry.[59]

The validity of considering the development of a technology as a so-called "public good" (something which, by virtue of its widespread potential benefits and great expense and/or risk, is developed through public support rather than through private investment) has been a matter of substantial debate for the last decade or so. In cases such as the SST and water and air pollution control, for example, where the expense and/or the uncertainties are enormous, it is extremely unlikely that private industry will be either willing or able to bear the burden of developing increasingly complex technology—especially when the potential benefits are undemonstrated.[60] This was clearly the case with FPC in the United States, even though its capital requirements were hardly overwhelming (total federal expenditures amounted to $14 million). Yet, despite the relatively modest capital costs, critics were able to question the value of or need for FPC in light of the adequate levels of protein available in the United States.

A number of Senators also testified before the House Subcommittee in support of the FPC legislation. Senator Bartlett (Alaska), one of the original proponents on the Senate side, and a representative of a fishing state, came prepared with dramatic photographic evidence of children in South America who had been fed FPC. Bartlett strongly urged the approval of a five-plant measure as what he termed an "international demonstration of good faith" on the part of the United States with regard to its intention to fully utilize its fisheries resource. Such a demonstration would be valuable, according to Bartlett, in the ongoing effort to extend the territorial limit from 3 to 12 miles in order to protect the U.S. fishing industry from the increasing encroachment of the Russians and Japanese.[61] Senator Edward Kennedy (Massachusetts), representing another fishing state, also testified in support of multiple FPC plants, to be constructed at the discretion of the Secretary of the Interior.[62]

Finally testimony, both pro and con, was presented by representatives of the food industry. While officials of the General Foods Corporation and others appeared in support of the legislation, the Secretary of the National Milk Producers Federation, E. M. Norton, argued against the program:

> While there might be some sound base for turning to questionable sources of supply for protein food in times of great emergency and shortage, there does not appear to be any valid reason for using matter ordinarily considered not suitable for food when we have on hand supplies of high-protein food (for example, nonfat dry milk) produced under high sanitary standards from acceptable sources . . . There is nothing new about the recovery of foods from material which is ordinarily considered unfit for food. This is merely a new application of the idea to fish, including the offal and waste . . . There is no need for fish flour made from inedible portions of fish, and, in the absence of a compelling emergency, its sale ought not to be permitted.[63]

From the perspective of the "world food crisis" of the 1970s, the short-sightedness of this statement is evident. However, it primarily reflects the views of an industry threatened by competition from FPC, rather than an informed commentary on world protein availability.

Final Congressional Compromise

Based on the hearings held by the Subcommittee on Fisheries and Wildlife Conservation in August, an FPC demonstration plant bill was reported to the House floor on September 30, 1966. The committee recommended passage of a companion bill to S. 2720 with the following differences: (1) only one plant, either constructed or leased, was to be authorized; and (2) FDA approval of FPC was to be obtained before a plant could be built or acquired.[64]

Senate members of the Conference Committee hoped to be able to retain the multiple-plant concept (derived from the water desalinization program) in the compromise version of the bill. House representatives were adamant, however, in their opposition to construction of more than one plant. A compromise was reached when John Pastore (Rhode Island), the second ranking Democrat on the Senate Commerce Committee, convinced the Conference Committee to agree to a second plant to be leased at Point Judith, Rhode Island.[65] Thus, one plant was to be constructed under an initial appropriation of $1 million, the amount recommended by the BCF; and one plant was to be leased under annual appropriation of $1,550,000 lasting for five years. This latter amount was to cover both the leasing of the Rhode Island facility and operation and maintenance costs for *both* plants.

The compromise was accepted and passed by both houses as P.L. 89-701, and subsequently signed into law by President Johnson on November 2, 1966. The plant, known officially as the Experiment and Demonstration Plant, (EDP), was to be designed with a 50 ton per day raw fish input capacity. The Secretary of the Interior was authorized to select the specific sites for both plants, to conduct any further R&D required, either through the BCF or by private contract, and to administer plant operation and maintenance (also through private contractors). Under the terms of the compromise agreement, no action could be taken until the FDA approved the domestic use of FPC.

PROGRAM IMPLEMENTATION (1966–1969): PRODUCTION AND TESTING DESPITE UNANSWERED QUESTIONS

Another FDA Obstacle

Despite congressional action on the "FPC bill," all possibilities for further progress still depended on the decision of the Food and Drug Administration. Although the Bureau had managed to allay FDA fears regarding the issues of IPA toxicity and lead residues in whole FPC (see chapter 4 for further details), no resolution had yet been achieved concerning the acceptable level of fluoride in the product. Now the BCF heard unofficially of yet another problem on the horizon. In response to the original FDA concern for the protection of the "unknowing consumer" (see chapter 5), the regulatory agency was considering the imposition of a bulk packaging restriction, when and if the product was finally approved. The object of such a regulation would be to limit sales of FPC to small, clearly labeled quantities and, therefore, to effectively prevent the use of the product in manufactured foods in which the consumer would be unaware of its presence.

By late fall it had become clear to the BCF leadership, which was anxious to move forward with the demonstration plant project, that the only way to avoid further delays over the fluoride issue would be to partially reduce the bone and scale content of the raw material. The VioBin Corporation had also come to a similar conclusion, and both petitioners submitted new samples to the FDA in November 1966, showing reduced fluoride contents. In view of the fact that FPC was being promoted for use in developing countries where concerns about fluorosis, if any existed, were insignificant in comparison to the need for protein, it is noteworthy that both the BCF and VioBin were willing to quietly comply with the conservative fluoride regulations of the FDA.[66] This concession was at least partially responsible for making FPC inaccessible to the target populations for which it was intended, the malnourished poor in the developing world, since decreasing the bone and scale content substantially increased the cost of FPC production relative to other protein supplements, such as soy flour, due to the need for additional equipment and energy.

The Lead Agency Concept

One possible source of leverage in convincing the FDA to approve whole FPC was the Marine Sciences Council. The MSC had received strong support from its member cabinet officers, especially Interior Secretary Udall, and was then under the vigorous leadership of Vice President Humphrey. During the early part of 1967 work on the FPC project had progressed through the initiative of Edward Wenk, Jr., Executive Director of the MSC, and former BCF Director, Donald McKernan, then with the State Department. A draft proposal was evolved that designated the Agency for International Development in the role of "lead agency"[67] for the funding and administrative execution of a program entitled "Food from the Sea for Undernourished People of the World." The program was to be implemented by the Nutrition Branch of AID's Office of Technical Cooperation and Research and established as the Food from the Sea Service in the Office of the War on Hunger. Accordingly, an interagency committee composed of representatives from AID and BCF was established to coordinate the project. In light of later program difficulties, it is significant that during the early stages of work by this committee there was, according to Edward Wenk, "never a single, negative signal from the BCF that there might be a mismatch between their enthusiasm for the program and their performance."[68]

The designation of AID as the lead agency was a source of some concern and debate among the members of the MSC. "On the one hand, the AID history included evidence of fluctuating leadership and policy, generally slow-moving management, and a running battle with Congress. On the other hand, the agency had the responsibility and experience for overseas technical assistance and was already staffed abroad by representatives who could bridge the gap to direct recipients."[69] AID also had its reservations about the Food from the Sea Program. For one thing, impetus for the effort was coming from a source external to the agency (the White House), a fact that continued to be an irritant to the agency throughout the course of the program.[70]

AID was faced with some other uncomfortable facts: (1) there was still no commercially viable production of FPC in the United States, only talk of production; (2) because there was no production, there were no reliable sources of data on which to estimate accurately the economics of FPC, especially in relation to other protein sources; (3) there were no known markets for FPC in terms of its use in fortified products; and (4) there was no information on the degree of market acceptance for FPC-fortified products.[71] Finally, AID was faced with the need to protect itself in the event that the FPC program failed due to the inability to win FDA approval.[72]

In light of these facts it is somewhat surprising that AID, in the role of "lead agency," ever got the FPC program off the ground. The lead agency concept was ill-fitted to the nature and mission of the AID organizational structure since the idea had originated in the Department of Defense where it referred to the development of system projects such as bomber or missile construction in which the DOD would take the lead role. AID, however, did not have the staff or the clout to coordinate the work of other agencies, and its original assignment as lead agency may be seen from this perspective to have stemmed from a misconception on the part of the MSC staff of the role of AID as a monolithic unit through which U.S. foreign aid was distributed. Because the development effort had come primarily from the USDI, which had promoted FPC to better utilize fishery resources, than from the State Department, which was more concerned with feeding hungry people in the developing world, there was little incentive for AID to pursue the use of the product for international food relief. Even after the joint AID-BCF program was established, sufficient funds were never allocated to the project in the USDI budget, and thus, with constant pressure from the Bureau of the Budget for budget tightening, department heads were much more likely to support projects that had originated within their own agency than projects imposed upon them by the White House.

The Congress, on the other hand, did not share AID's reservation about the program. After approving the construction of the Experiment and Demonstration Plant (EDP) in the closing days of the 89th Congress, the newly elected 90th Congress further endorsed the use of FPC by amending the Foreign Assistance Act of 1961 to give the President the authority to demonstrate the potential and encourage the use of fish and other protein concentrates to reduce nutritional deficiencies in less developed countries.[73] The action by Congress closely paralleled the principal recommendation of the first report to the President and Congress by the Marine Sciences Council. This summary document proposed an intensified long-range program to exploit the oceans as a source of food, using FPC as a first step, "to help feed the undernourished people of the world."[74] The humanitarian theme, which had been used to justify the FPC program since its inception in 1961, remained as one of the two principal motivating rationales until the end of the Johnson Administration in 1968.

FDA Approval of Whole FPC

The exact combination of factors that eventually convinced the FDA finally to approve FPC as a food additive is not clear. Certainly the steady pressure exerted on the HEW Secretary John Gardner and on the FDA Commissioner James Goddard[75] by Vice President Humphrey, Secretary Udall, and members of Congress had some effect, as did the creation of the Food from the Sea Service within AID. But the critical factor may well have been the results of the crash fluoride testing program conducted by the BCF (at the suggestion of the FDA), which were submitted to the FDA on January 4, 1967. The results indicated that, if a 20 gram per day FPC intake were assumed, an average child would ingest 1.6 milligrams of fluoride per day. In the view of the NAS Subcommittee on Marine Protein Resource Development such a level would not cause objectionable mottling of teeth. This was also well within the safe limit of 3 to 5 milligrams per day set by the FDA.

Thus, after five years of testing, negotiations, and further testing by the BCF, the Food and Drug Administration announced on February 2, 1967, the tentative approval of the BCF food additive petition for the manufacture of whole FPC using fish of the hake family. In the process of sanctioning the use of FPC, however, the regulatory body imposed several significant restrictions.

IPA residue levels were limited to 250 ppm.

Ethylene dichloride residues (used in the VioBin process) were limited to 5 ppm.

Fluoride levels were set at 100 ppm.

Total FPC in the diets of children up to eight years old was not to exceed 20 grams per day.

FPC could only be sold directly to the consumer in clearly labeled, one-pound (or less) packages.

In the preamble to the ruling the FDA further disclosed its rationale for the ban on bulk sales and consequent use in manufactured foods: "Individual consumers are entitled to the opportunity of choice in the matter of whether they desire to include this item in their diets . . . Bulk sales or packaging in units larger than one pound cannot be authorized unless the Commissioner is supplied with data demonstrating that the proposed use of the article will promote honesty and fair dealing in the interest of the consumer."[76] Once again a vestige of the early filth issue had manifested its power over the thinking of the FDA. In the words of one official, the one-pound package limitation was imposed "as a means of dealing with the substantial number of consumers who do not wish to eat a product that contains viscera and intestinal contents."[77]

The FDA order provided that anyone adversely affected could file an objection within thirty days. Over 125 such objections were received (see, for example, exhibits 1 and 2, chapter 4), more than 90 percent of which were from various dairy interests.[78] The dairy industry position was that it was forced to spend millions of dollars a year to maintain rigid sanitary conditions (often enforced by as many as fifteen annual inspections per plant). In its view, that approval of FPC would constitute a special exception to the code of standards for pure food and would therefore represent unfair competition. In addition, the dairy interests feared the widescale distribution of FPC would seriously undermine the low-cost world protein market which had been dominated by nonfat dry milk. At that time a surplus of dry milk existed, and FPC was viewed as merely exacerbating an already poor economic environment.

The milk industry was joined by other industry representatives such as the wheat flour interests who were concerned that continued FDA scrutiny of the contents of FPC might well subject the contents of their products to further investigation. It thus appeared that a major objection would be lodged against the FDA ruling, necessitating a formal, public hearing. But, at the last moment, the major dairy objector, the American Dry Milk Institute, withdrew its filed objection citing the increasing world food deficit highlighted at the February 1968 World Food Conference in Rome.[79] Apparently, this change of heart by the Institute also reflected some effective interest group bargaining by then Vice President Humphrey regarding changes in federal milk regulations (see chapter 4 for further details).

Experiment and Demonstration Plant: The Point of Decision

With signs pointing to the fact that the FDA ruling would not be seriously contested after all, the BCF had to decide whether to proceed immediately with its Experiment and Demonstration Plant program, which was already more than six months behind schedule, or continue to await formal FDA approval. In view of the support shown by the President and the Vice President, the Marine Science Council, and the fishing industry, the BCF leadership decided that it would be politically unrealistic to delay the construction until after the FDA ruling became final. Despite the risk of much second-guessing on the part of Congress and the public in the event that the FDA failed to approve the product, the judgment was made to proceed on the assumption that the FDA ruling would be sustained.

Crucial decisions had already been made within the Bureau about the location of the two plants. Due to the earlier Conference Committee compromise with Senator John Pastore, the leased plant was to be located at Point Judith, Rhode Island. The selection of the site for the plant to be constructed, however, was of a slightly more complex nature. There was never any doubt that, with many of the major congressional proponents coming from the State of Washington, for example, Senator Warren Magnuson and Congresswoman Julia Butler Hansen, the plant would be located somewhere in that state. The question was only which one of a number of potential sites would be chosen.

Because Representative Hansen, as chairperson of the Interior Appropriations Subcommittee, had been personally active and influential in guiding the original EDP bill through the Congress, she strongly urged that the location of the plant be in her district, specifically at Aberdeen, Washington, on Gray's Harbor of Puget Sound. This site was considered acceptable by the BCF since it provided access to both "inside" stocks (within Puget Sound) and "outside" stocks (in the Pacific Ocean).[80] It did, however, present Senator Magnuson with a difficult political problem. All the representatives of the other potential locations, also predominantly on Puget Sound, were his constituents and, inevitably, most would not be pleased by the announced choice. But the Pacific hake resource appeared at the time to be available at Aberdeen, and the Gray's Harbor Development Authority had lobbied actively on behalf of the location. Eventually, agreement on Aberdeen as the site for the EDP was obtained. As with so many of the decisions affecting FPC, it is clear that the site selection was made on the basis of political realities rather than on accurate engineering data. Later, the growing scarcity of fatty fish resources (sardines) off the Washington coast would damage the Bureau's attempt to have the FPC program continued.

By May 1967, a Request for Proposals (RFP) was published for the design, construction, and operation of the EDP. Seven proposals were received by the BCF although four of these did not meet the basic quality standards established and were immediately rejected. Three others, those from Ocean Harvesters, Inc., Aqua Protein, Inc., and Stearn-Rogers, Inc., were accepted for consideration; and a fourth, from Resources Engineering, Inc., was subsequently accepted with proposed revisions.

One fact was immediately apparent from the proposal bids received: it would be impossible to construct the EDP for the $1 million authorized under the terms of PL 89-701. The bidders cited the delays in receiving FDA approval (thus driving up the construction costs due to inflation); the necessary improvements in design standards caused by the increased stringency of the recent FDA restrictions; as well as the increased costs of government inspections. It is also apparent that part of the difficulty was inherent in the terms of the enabling legislation itself which attempted to specify a raw fish capacity of 50 tons per day as well as a total cost of $1 million, thereby limiting severely the design of the bidders' proposals. Eventually, however, the $1.9 million Ocean Harvesters[81] proposal was selected, and, when the hoped-for commitment of private capital[82] did not materialize, it became clear that Congress would have to be approached for additional funding.

By the time that the need for supplemental congressional appropriation had become unavoidable, the leadership at the Bureau had changed with Harold Crowther replacing Donald McKernan as director. Crowther maintained serious reservations about the continued viability of the program that were only magnified by the EDP funding difficulties. Apparently, some consideration by the BCF management was actually given at this point to terminating or substantially curtailing the FPC program. However, with many jobs hanging in the balance and with the lack of available hard evidence against the program to present to congressional proponents such as Hansen and Magnuson, the FPC development seemed to have taken on a momentum of its own and no restrictive decisions were made.[83]

This suggests the more general difficulty within the administrative process of attempting to "turn a program off" once it has begun. Due to the incremental nature of decision making in the political process it is generally too risky, both personally and politically, for a bureaucrat to ask the really hard questions about a program in difficulty within his agency. When his and other people's jobs stand in jeopardy, it is easier and safer to adopt a low profile position until and unless the program should come under the scrutiny either of the White House (for example, through the OMB) or the Congress (through the GAO). This is not to suggest that there were no individuals within the BCF and elsewhere in the bureaucracy who were actively engaged in questioning various aspects of the FPC program. There were many. But the facts indicate that the suggestions of these individuals were either consistently ignored (or countermanded) or that, at crucial decision points, their recommendations were simply overlooked in the name of political expediency.

AID—Food from the Sea Program

While the BCF program was delayed pending the resolution of its funding problems, the Food from the Sea Program was moving forward. The interagency group coordinated by the MSC had produced a proposal to identify one of these countries for a model FPC program. The intention of the model program was to promote indigenous processing and distribution consistent with local needs, fish supplies, and customs. The purposes of the overall AID-FPC program plan can be summarized as follows:

Develop in the United States a commercial process for producing FPC including the design, construction, and leasing of pilot plants as authorized by PL 89-701.

Improve the fish catching, landing, and processing capabilities of three protein-deficient countries.

Develop markets for FPC in at least one protein-deficient country.

Establish a viable commercial FPC system in at least one protein-deficient country.

Encourage other nations and private interests to establish commercial fishing industries wherever feasible.[84]

In pursuit of these objectives AID sent out interdisciplinary survey groups to sixteen less developed countries during 1967–1968 to determine which LDCs might provide the best FPC opportunities.[85] A twenty-point checklist was prepared for use by the team in each country. The primary purpose of that list was to answer the following questions: (1) Was there an adequate supply of fish available (at least 20,000 tons per year) at reasonable price (no more than $20 per ton)? (2) Were the local foods used by low-income people suitable for fortification? (3) Were the local food processors willing and able to market FPC-fortified foods? No one country met all of the criteria established by the Food from the Sea Program, but Chile, South Korea, and Morocco were selected as the three LDCs having the greatest potential for success and thus, as the sites where feasibility studies would be conducted. In retrospect, it is likely that the countries chosen did not necessarily represent the three with the greatest potential for FPC production.[86] Although both Chile and Morocco were justifiable on the basis of resource and prior interest, the rejection of Peru, with its vast anchovetta resource, was, it would seem, a political decision based on the harassment of the U.S. tuna fleet by the Peruvian government.

A Participating Agency Service Agreement (PASA) was signed on April 7, 1967, by Herbert Waters, representing the AID Office of the War on Hunger, and Harold Crowther, Director of the BCF, providing $200,000 of AID funds to continue FPC research at the College Park Laboratory and elsewhere.[87] AID also announced plans to make an additional $1.8 million available during fiscal 1969 for the three-country feasibility studies.

By January, the Food from the Sea Program office staff was complete,[88] and, at the end of that month, the first Food from the Sea industry conference was sponsored by AID. Its objective was to acquaint U.S. industry with the government's program and to provide a foundation for industrial participation in the AID initiative by contracting for the commercial production of whole FPC to be used in experimental overseas feeding projects.

At the conference AID officials presented the preliminary results of the sixteen-country FPC studies to about fifty representatives of thirty private firms.[89] Despite the conference effort, however, many companies remained confused and hesitant as to the profitability of producing FPC, especially with regard to its use in LDCs. As a result, only two firms responded to the Request for Proposals (RFP) offered by AID for the commercial manufacture of the product: Alpine Marine Protein Industries of New Bedford, Massachusetts (a subsidiary of Alpine Geophysical Associates, Inc., of New Jersey and a licensee of VioBin Corporation) and the Astra Group of Sweden.[90] The Astra Group subsequently dropped out of the bidding, however, because of the AID requirement that the plant site and labor supply be in the United States, thus leaving only the Alpine Marine Protein Industries proposal.

The Alpine Corporation had acquired expertise in the field of FPC manufacture by purchasing the production rights to the VioBin process in 1967.[91] Their bid (based on the AID limits of $900,000 for the product and $10,000 for shipping) was for 1000 tons of FPC at $.42 per pound, with one-third to be delivered by January 1969 and the remainder by the following July. The proposal was accepted and a contract signed on April 26, 1968. The Bureau of the Budget at that time was advocating the use of the so-called "performance contract," an agreement whereby the contractor is paid after delivery on the basis of the product fulfilling previously agreed-upon standards. This kind of contract was considered a good tool for controlling quality of output on new and experimental technologies. Because of subsequent difficulties with the product produced by Alpine, the existence of a performance contract became important, especially in light of the fact that BCF personnel had stated their serious reservations to AID about the ability of Alpine to fulfill the terms of the contract.[92]

In fact, a number of other issues may be raised about the contractual agreement with Alpine. Should AID have proceeded with the signing of a contract when only one serious bidder remained? It has been suggested by certain AID officials that the Agency was left with little choice because of the major delays in the construction of the demonstration plant by the BCF. But this also leads to a broader question: in view of the continued reluctance of private industry to become involved (as witnessed by their failure to bid on the AID RFP), why did AID pursue private, domestic production when their primary mandate was to develop FPC production capabilities in developing countries? The agency certainly would have been fully justified in terminating the program instead. The less than satisfactory answer that AID was under continuing White House and Congressional pressure to produce results and needed the product to proceed with the other aspects of the program does not appear to be supported by the evidence at hand. In fact, it would appear that there were other even less acceptable alternatives, such as dissolution and transfer of the program or congressional inquiries, that the Food from the Sea Office was trying to avoid.

AID had also, by this time, prepared and disseminated guidelines to the American Voluntary Agencies and AID Overseas Missions for the use and evaluation of FPC in pilot feeding programs which resulted in a highly favorable response—over thirty proposals were received requesting 1300 tons of FPC for trial use. The desire to fill these requests provided yet another reason for the agency to press on with domestic production. By the end of the spring of 1968 AID was prepared to award contracts totaling $750,000 for the in-depth feasibility studies of the three countries selected to participate in the FPC program, Korea, Chile, and Morocco. Subsequently, it was decided to hold the Morocco study in abeyance,[93] and the field was narrowed to five proposals on the remaining two countries. In June, General Oceanology, Inc., of Cambridge, Massachusetts, was awarded the contract for feasibility studies in Chile and Korea. The analysis was to consist of three parts: (1) market analysis, to determine if FPC-fortified food would be acceptable to local populations; (2) supply analysis, to determine the availability of an inexpensive and underutilized source of fish and the relationship of cost to demand; and (3) product development and testing studies, to involve flavor and product acceptability tests.

National Center for FPC Established

In spite of the uncertainties over funding for the EDP the Bureau was in the process of restructuring its domestic program by establishing a National Center for Fish Protein Concentrate, which would combine all FPC R&D activities ongoing within the BCF. Roland A. Finch was appointed as the first director of the center in 1967, with dual responsibilities to conduct FPC research through the College Park Laboratory and to administer the contract for the EDP as Technical Officer. The continuing importance attached to the

FPC program by the Johnson Administration was underscored in the President's message to the Congress of February 8, 1968, entitled "To Build the Peace":

> We must also tap the vast storehouse of food in the oceans which cover three-fourths of the earth's surface. I have directed the Administrator of the Agency for International Development and the Secretary of the Interior to launch a 5-year program to
> 1. perfect low-cost commercial processes for the production of FPC;
> 2. develop new protein-rich products that will fit in a variety of local diets;
> 3. encourage private investment in FPC production and marketing, as well as better fishing methods; and
> 4. use this new production in our "food for freedom" program to fortify the diets of children and nursing mothers.[94]

But unfortunately, besides the pursuit of research at the National Center for FPC, the BCF could do little more in response to the President's directive until the Congress acted on its request for a supplemental appropriation to facilitate the construction of the EDP.

Congressional Action on Funding (1968)

The long months of waiting, since the realization that the EDP could not be constructed with the amount appropriated, finally came to an end for the BCF on March 26, 1968, when a new set of hearings was initiated before the House Subcommittee on Fisheries and Wildlife Conservation on H.R. 15490 and 15535 (related to S. 3030), amending the original enabling legislation, PL 89-701. Under the chairmanship of John Dingell, representatives of the Bureau and other witnesses faced some severe and understandably skeptical questioning from members of the subcommittee.

BCF Director Harold E. Crowther led off the testimony by stating the reasons that $1 million had been insufficient to build the plant: (1) the preliminary cost estimates had been developed based on limited engineering data available in mid-1965, and since then the need for a five-stage extraction (rather than three) had become apparent; (2) the tendency of experimental prototype facilities to cost more than commercial production facilities of the same (or even larger) capacity, due to built-in requirements for sanitation, research capability, flexibility, and instrumentation, all of which make the plant more expensive; (3) the final FDA specifications concerning fluoride were not known at the time of the final design; and (4) construction costs had risen dramatically since 1965.[95]

Similar testimony by Howard Wright of Ocean Harvesters, Inc., the construction and operation contractors, revealed that the original $1 million figure had been derived simply by reducing the original $5 million for five plants to an equivalent figure for one plant. But, as Wright noted, "the scientific and economic objectives of this one plant embrace many of the objectives of the original five plants."[96] In addition to inquiring about the manner in which the costs of constructing the EDP had been determined, Representative Joseph Karth (Minnesota) demanded an explanation of the method of contracting which had been used. He suggested that the use of the cost-plus-fixed-fee contracting system, where the contractor is guaranteed his cost plus an agreed-upon bonus fee, was not appropriate for a facility of this sort and that a majority of similar projects were let on a purely competitive basis.

Other individuals testifying at the hearing were concerned about competition of another sort, namely, competition between the government and private industry. Representative William Springer of Illinois and his constituent, Ezra Levin, the President of the VioBin Corporation, reintroduced the question of whether the government should be involved in something already developed by the private sector.[97] But the testimony of Levin and Springer was rebutted by a number of the later witnesses. Director Wenk of the Marine Sciences Council reemphasized the fact that the majority of private industry would not become involved due to the absence of a domestic market and the unreliability of foreign demand. Howard Wright, President of Ocean Harvesters, asserted that he knew of no process with a conclusively demonstrated commercial capacity.[98]

After further probing by Chairman Dingell and other subcommittee members,[99] debate focused on the most efficient means to fill the EDP funding deficit. Wenk and Crowther had reported previously that the second, leased plant was never pursued because, in the first place, a properly sized and suitably flexible facility could not be identified[100] and, secondly, funds for the second plant, although authorized, were never released by the Bureau of the Budget.[101] On this basis Congressman Alton Lennon (North Carolina) proposed that the annual appropriation of $1,550,000, which had been authorized for the leasing of the second plant and operations and maintenance at both plants, be used to make up the funding deficit. His recommendation was endorsed by the subcommittee, and a new law reducing the number of experimental plants (PL 90-549, which amended PL 89-701) was subsequently approved by Congress and signed by President Johnson on October 4, 1968.

A number of points should be made about the outcome of the 1968 hearings. First, there is little doubt that the luster of the FPC program had been diminished in the eyes of Congress by the need for a supplemental appropriation. The suspicion of incompetence due to the constant shuffling of cost figures and time requirements by Bureau officials hung over the BCF-congressional relationship for the remainder of the program's life. The Bureau may also have sealed its own fate by agreeing to the Lennon compromise of using the funds earmarked for the leased plant (to be built on the East Coast) to supply the deficit for the plant being constructed at Aberdeen, Washington.

By giving up the flexibility of a plant on either coast, the BCF was, in effect, putting all of its eggs into one basket, thus becoming solely dependent on the continued adequacy of the west coast fisheries resource.

There were also clear indications at the 1968 hearings of a shift in the BCF policy regarding the primary objective of the program. Delays in the construction of the EDP had so far prevented the Bureau's involvement in the AID overseas feeding program, which was the original humanitarian rationale for development. The VioBin-licensed process used by Alpine appeared capable at that time of meeting AID's needs without the help of the Aberdeen facility. Political justification for the program at higher policy levels thus began to shift away from feeding the "starving millions" toward the notion of aiding the U.S. fishing industry, which had continued to suffer financial difficulty through most of the decade.

The Bulk Packaging Restriction

With the removal of the funding obstacle to the implementation of the EDP, attention in both the public and private sectors turned once again to the problems posed by the 1967 FDA regulations. The major issue remaining with the FDA was, of course, the bulk packaging restriction, which effectively prevented any serious entry by the food industry into the FPC market since the product could not be used in manufactured foods. Alpine Marine Protein Industries, the company under contract to AID, found the ruling particularly objectionable since it limited sale of their product only to AID for use overseas. In late July, the company formally petitioned the FDA for removal of both the one-pound limitation and the restrictions on the use of the product by children less than eight years old.[102] On August 16, the FDA replied to the petition, finding it inadequate due to the lack of data substantiating the effects of the proposed modification.

The effort to elicit a relaxation in the FDA regulations continued throughout the remainder of 1968 (see chapter 4 for details) although the Alpine corporation eventually opted to give up the fight, at least temporarily. But the matter had aroused the concern of the NAS Food and Nutrition Board which adopted a resolution strongly urging the removal of packaging restrictions on whole FPC since they were preventing the marketing of FPC in the United States and had no scientific justification. The resolution, formally transmitted to the FDA by the President of the NAS, Dr. Frederick Seitz, was received in a more conciliatory climate under the leadership of the new FDA Commissioner, Dr. Herbert Ley.[103] In response to Seitz's letter Ley noted that the FDA "recognized substantial interest in having the restrictions removed from the present food additive regulations."[104]

PROGRAM DECAY (1969–1972): DEVELOPMENT IN THE FACE OF TECHNOLOGICAL UNCERTAINTY

EDP Construction

Construction plans had been moving forward for the experiment and demonstration plant (EDP) in Aberdeen, Washington, with the design based on the use of the lean fish, Pacific red hake, as the design standard. After receiving the necessary certification[105] from HEW Secretary Wilbur Cohen in March 1968, a final contract was awarded to Ocean Harvesters, Inc., the following October for the design, construction, and operation of the plant. By March 1969, a specific plant site had been acquired at Aberdeen and a deed transferred by the Gray's Harbor Port Authority.

During the summer of 1969, in-house (within the BCF National Center for FPC) examination of the cost figures submitted by Ocean Harvesters, Inc., revealed that, despite the supplemental funds authorized by Congress in 1968, the plant might still cost more than the funds available would support. Staff of the National Center for FPC at College Park became concerned that, in order to operate within the statutory funding limit set by Congress, it might become necessary to eliminate and/or undersize the equipment. This would have the effect of removing much or all of the flexibility from the EDP, thereby eliminating the experimental function of the plant. The staff was also worried that cost-cutting changes in the system design would cause serious practical and political problems as well, for example, with the disposal of 1700 gallons of waste per hour containing fish oil, IPA, and other waste products.[106]

The College Park group asked that the BCF director be informed that

> the consequence of capital cost limitations is a drastic increase in operating costs, because of increased personnel requirements. As a result of this imbalance between capital and operating costs, the plant will *not* demonstrate the commercial feasibility of manufacturing FPC because it will not be possible to extrapolate cost figures to a commercial plant with adequate precision.[107]

This information was related to BCF director Harold Crowther, who responded by indicating his understanding of the consequences of the proposed revisions. However, he reiterated that, because of statutory restrictions imposed on the BCF by the terms of the legislation, the Bureau had little choice but to move ahead within the limits of the funds allocated. Consequently, design work continued as originally planned, and, in the fall of 1969, the BCF began the acquisition of long lead-time equipment and the modular fabrication of processing equipment.

AID Program Setbacks

The AID program had also not been without its difficulties. Alpine Marine Protein Industries, which had contracted with AID to produce 1000 tons of FPC by July 1969, had been forced to request an extension on the delivery date until November because of an inadequate supply of red hake. To make matters worse, in June 1968 two staff members within the Food from the Sea Program had departed and their positions were not filled.[108] But the most serious blow was yet to come: upon the delivery of the Alpine-produced FPC in November 1969 it was discovered that the product did not meet the protein efficiency ratio (PER) specifications required by the FDA.[109]

Out of 1000 tons originally contracted for, Alpine delivered 426 tons, of which only 172 were accepted by AID as meeting PER specifications. On the basis of the faulty product quality and the delays in delivery, AID was compelled to terminate the Alpine contract on November 26, 1969. The delivery delays were blamed by Alpine on production problems resulting from the unavailability of raw material at a reasonable price.[110] Although it was true that the fish catch that season was unusually poor, there were additional problems with the product relating to milling quality and bacteriological count. The issue of the PER specification became the center of a controversy and legal confrontation between Alpine Marine Protein Industries, which contended that the VioBin-licensed process was at fault, and the VioBin Corporation, which argued that the process had been used improperly.

In any event, with less than 20 percent of the amount of FPC originally anticipated, the distribution plan developed by the Food from the Sea staff required total revision. AID had planned initially to work closely with the American voluntary agencies such as CARE to determine the most effective ways of utilizing FPC under various cultural conditions. Due to the overwhelming response (thirty proposals requesting approximately 1300 tons of FPC) that AID had received, the agency had even contracted with the University of California at Davis to provide expert assistance in determining which proposals were best suited to the goals of the program.

Now, however, with only 172 tons of FPC available (of which an additional 33 tons were later rejected due to substandard protein levels revealed in bioassays), all of the proposals for test feeding were open to reconsideration. Contingency plans were hurriedly created in an attempt to prop up the FPC program until a suitable product could be made available from the BCF plant under construction on the West Coast. In the meantime it was decided to take the remaining 139 tons and to offer 100 tons to Chile, 36 tons to Brazil, and 3 tons to Thailand. Interestingly, none of the product was allocated to the other two countries involved in the feasibility study, Korea and Morocco. The product was accepted and used in Brazil and Thailand, but held in Chile where it was subjected to intensive bioassay.[111]

Pressure was subsequently brought to bear on AID and the Chileans by Ezra Levin (through an exchange of correspondence with George Parman of AID and Dr. Julio Santa Maria of the National Health Service of Chile) to accept the FPC with the lower PER values. Food from the Sea Director Parman replied to Levin, however, that besides the fact that AID was legally precluded from accepting a product that did not meet contractual specifications, there were also "serious practical and political risks." He referred specifically to the adverse publicity for the United States if it became known that AID was shipping products such as FPC for use overseas that did not meet specifications for use in the United States.[112]

By early 1970, the entire Food from the Sea Service had run into serious political and bureaucratic problems under the changed priorities of the new Nixon administration which had taken office in January 1969. Despite the continued support of the Marine Sciences Council (although now without the strong and enthusiastic leadership of Vice President Humphrey), there was little interest in or favor for such AID efforts under the fiscally conservative Nixon administration. A long internal agency power struggle had resulted in the termination of the War on Hunger office, of which the Food from the Sea Service had been a part, and the transfer of the fisheries program to the Technical Assistance Bureau under the Office of Agriculture and Fisheries. The fisheries staff under this new arrangement was severely reduced and consisted essentially of only two professionals.

The termination of the AID–Alpine contract thus was the initial event in a process that ultimately resulted in the total phase-out of AID involvement with FPC. The decision to withdraw AID participation had a number of other consequences for the FPC development effort. First, it cut off the most important and natural outlet for the product, namely feeding programs in the developing countries. Second, as a result of the loss of overseas distribution, the entire rationale for the program as a nutrition intervention to combat malnutrition was discredited. Finally, the loss of this important humanitarian justification, compounded by the negative record of the program, served to further jaundice the already skeptical perception of FPC in the minds of key congressional leaders.

National Marine Fisheries Service Established

In 1969, Harold Crowther departed as the director of the BCF, leaving a serious management vacuum which was filled by a temporary three-man "troika" arrangement. At about the same time, Secretary of the Interior Walter Hickel fell out of favor with President Nixon. The Office of Management and Budget (OMB), formerly called the Bureau of the Budget, was consequently directed by the White House to recommend cuts in the USDI budged. An OMB official released a letter shortly thereafter stating that the

BCF should not be doing for the fishing industry what the industry was capable of doing for itself.[113] OMB suggested instead that if programs such as FPC were really that crucial then their funds should come out of the departmental operating budget. Such a proposal gained the Bureau little popularity within the USDI.[114]

Other reorganization actions by the Nixon administration were also affecting the Bureau of Commercial Fisheries. Under the terms of the Reorganization Plan No. 4 of October 1970 the BCF was shifted from the Department of the Interior to the Department of Commerce where it was retitled the National Marine Fisheries Service (NMFS) as one component of the National Oceanographic and Atmospheric Administration (NOAA). For the commercial fisheries agency, the new relationship proved very beneficial since the promotion of commercial fisheries development lay much closer to the mainstream of the Commerce Department's interests in a healthy fishing industry than it did to the Department of the Interior's range of concerns (in the use of fishery resources).

Thus, it does not appear that the shift of the Bureau to the Department of Commerce had any significant effect, either positive or negative, on the operation of the FPC program. The National Center for Fish Protein Concentrate remained intact through the transfer, although a number of personnel changes occurred as a result of the move. Under the aegis of the new NMFS, the FPC program was directed toward a variety of objectives including the demonstration and evaluation of the IPA extraction process at the semi-industrial scale and the commercial feasibility of FPC in terms of resources, economics, and markets.

FDA Regulations Modified

In December 1969, the problems and potential of FPC had come to the attention of the White House Conference on Food, Nutrition, and Health, which had been convened in recognition of the fact that even the United States had a nutrition problem in certain socioeconomic groups. The conference produced a report the following December that included a series of policy recommendations concerning the role of the federal government in the area of nutrition. One conclusion in particular stated: "The FDA should permit the addition of FPC in formulated food and Section 121.1202 of the Food, Drug and Cosmetic Act [*sic*] should be amended to permit such use."[115] At a meeting on this subject between NMFS and FDA officials that same month the regulatory agency suggested more study was still required on the processing and end product before the packaging restriction could be lifted.

The food and drug agency had actually amended the Code of Federal Regulations (Section 121.1202) on August 4, 1970, expanding the approved species list to include herring and menhaden using the IPA solvent extraction method only, not the VioBin dual-solvent extraction process.[116] But other issues were still to cause unforseen delays in FDA action on the bulk packaging issue. In early 1971 concern was expressed by staff members at the National Center for FPC about the concentrations of mercury and other trace elements occurring in the fish used to manufacture FPC as a result of industrial pollution. NMFS-sponsored studies were hastily initiated to counter anticipated FDA objections.

EDP Operational Problems

Progress on the EDP in Aberdeen, Washington, continued despite the uncertainties about whether private industry would ever be permitted to use the product in manufactured foods. Ground had been broken for the plant at Gray's Harbor on January 31, 1970, and construction of plant buildings began shortly thereafter. By October 1970, the plant processing equipment had been prefabricated in modular form and transported from SWECO, Inc., in Los Angeles, California, to the plant site in Washington. At the beginning of March 1971 construction had been completed and preliminary start-up operations begun.

On April 17, 1971, the plant was formally dedicated in a ceremony attended by 400 people, including two of the principal congressional advocates, Congresswoman Julia Butler Hansen and Senator Warren Magnuson. But, by the time of the official plant dedication, two years had elapsed since the passage of the supplemental funds to the original EDP bill, PL 89-701. To those responsible for the operation of the program it was becoming increasingly apparent that it would be impossible to achieve the intended goals of the program within the time frame authorized by Congress (which had a deadline of June 1972).[117] There were also additional financial woes caused by the reprogramming of between $1 and $1.5 million of NMFS operating funds, which were originally targeted for use in monitoring the plant operation and in helping to iron out the start-up problems, for use in a marine resources mapping project. As a result, the contract for design and construction of the EDP had to be rewritten to include a requirement that the contractor complete certain aspects of the necessary research that the NMFS scientists had been precluded from finishing.[118]

After a month of EDP operation, problems developed with waste disposal. The original Ocean Harvesters design had been predicated upon the existence of a nearby fish meal plant, to be operated by a local company known as Pacific Protein, Inc., which was to utilize the waste streams of fish from the EDP. But by October 1970, it had become known that the Pacific Protein plant would not be operated. This left Ocean Harvesters with the choice of dumping all of the effluent into a nearby river and municipal sewers or of using back-up incinerators. As of May 1971, the Environmental Protection

Agency standards had been tightened to such an extent that even the alternative of incinerating had been foreclosed.[119]

The waste disposal problem had been controlled sufficiently by June 1 to permit the first continuous extraction of Pacific hake under operating conditions. By the middle of that month the plant had actually reached baseline conditions in that FPC, meeting all FDA specifications, was produced under regular operating procedures. These achievements sufficiently impressed the visiting members of the NAS committee to the extent that they made two recommendations: (1) as a result of their concern about the disposition of the finished product which might enter commercial channels and compete with food already on the market, the committee recommended that only limited allotments be made available to the domestic food industry and to research institutions; (2) the committee proposed the convening of an international conference on the uses and potential of FPC.

AID Policy Revision

At about the same time that the EDP reached baseline operations, the AID intra-agency policy committee was meeting to formulate policy for future agency involvement with the FPC program. The committee had in hand the General Oceanology consultant reports on the viability of FPC in two of the three target countries, Korea and Chile. Concerning Korea the consultants had found that protein malnutrition existed but that present market economics and raw material shortages (the Korean annual fish catch had been grossly overestimated) militated against a commercial FPC operation at that time. They also found that up to 30 percent of the total Korean fish catch was being wasted due to lack of refrigeration and processing facilities. The report on Chile, on the other hand, was generally positive, although a commercial FPC venture was still considered to be a high risk unless the government of Chile guaranteed to purchase the product for use in national feeding programs.[120]

Even with the necessary amount of FDA-approved product being produced at the Aberdeen plant and a promising report from Chile, the advisory committee was reluctant to recommend proceeding with the program on the scale originally specified. AID had always considered FPC to be an "externally dictated project,"[121] and, for an organization used to dealing in proved, noncontroversial commodities, such as wheat and corn, FPC had been a source of almost constant trouble and frustration. Thus, the recommendation of the advisory committee was that, while the FPC program should be continued, it should be consolidated and carried out "in relation to total protein resources available for overseas feeding programs."[122] By the following year (1972) the FPC program had been relegated to a supporting role within the Office of Nutrition.[123]

In view of the fact that AID had had neither sufficient personnel nor adequate technical expertise to gauge accurately the real potential of the product, the inclination of the agency to back out of its "lead agency" responsibility was almost entirely predictable. The desire to be free of the FPC albatross was further reinforced by the pervasive feeling within the organization that the program had been imposed upon it by the White House through the MSC. With a new administration under Richard Nixon in office, and a correspondingly new set of foreign policy and development priorities, AID was only too willing to give up the project.

Demise of the MSC

It is not unlikely that one of the principal causes for the loss of commitment to the FPC program within AID was the related demise on April 30, 1971, of the National Council on Marine Resources and Engineering Development. There had never been great admiration within the budget-conscious Nixon administration for the activist role of the MSC in promoting marine affairs, or for that matter, for the whole notion of specialized agencies within the White House. In addition, unlike Hubert Humphrey, Vice President Spiro Agnew took virtually no interest in the MSC and generally did not even attend its meetings. When the statutory limitation of the council was reached, Congress was not requested to renew it.

The MSC clearly had a significant impact on the FPC program. The council had been primarily responsible for the high degree of visibility and interagency coordination that had been developed. The council had also taken the initiative in the creation of a systematic budgetary and program planning process. Exhibit 2 illustrates the relationship between projected and actual FPC budget goals.

Advocacy of a program such as FPC development was, of course, inherent

Exhibit 2
Marine Sciences Council Budget vs. Actual Funds Authorized

FY	MSC Budget Goals for FPC	Actual Authorization	MSC Cumulative Total	Authorized Cumulative Total
1967	$ 600,000	$ 689,000	$ 600,000	$ 689,000
1968	1,209,000	2,129,000*	1,809,000	2,818,000
1969	2,554,000	1,855,000	4,363,000	4,673,000
1970	3,680,000	2,903,000	8,043,000	7,576,000
1971	2,900,000	2,138,000	10,943,000	9,714,000

Source: Northrop Service Report, No. I, pp. 2-38.
*Includes $1 million authorized for construction of the EDP, programmed over FY 1969 and 1970.

in the MSC mandate.[124] However, the strong leadership role assumed by the Council through the latter part of the decade proved beneficial as well as deleterious to the FPC program. On the one hand, the favorable publicity and bureaucratic coordination generated by the MSC contributed greatly to the promotion of the concept in the eyes of Congress and the other federal agencies. At the same time, the council's strong advocacy—perhaps rooted in the need to maintain favor and visibility in the eyes of the Johnson Administration—certainly contributed to the pressure on the BCF for premature development of the process at the industrial works scale.

It was the MSC that had originally sought to involve AID as the so-called "lead agency," which led subsequently to criticism of the council for its leading, rather than supporting, role with regard to FPC. In view of the problems that ensued from designating AID to run a program for which it was ill-equipped, as well as from the vigorous MSC advocacy of the Food from the Sea Program (both by director of the council Edward Wenk and by chairman of the council Hubert Humphrey), it is certainly conceivable that the BCF development program and the AID overseas program were pushed into the implementation stage before either agency was really prepared to carry out its particular mission.

Role of the Private Sector

With the dissolution of the Marine Sciences Council, the scaling down and consolidation of the FPC programs within AID, and the imminent completion of the NMFS demonstration plant, a new and conceivably final chapter in the U.S. FPC story seemed to be opening. Significantly, one factor whose presence had always been assumed in terms of the final development stage had never materialized, namely, the involvement of venture capital from the private sector.[125]

Perhaps the brightest hope for involvement of the major food producers lay with the General Foods Corporation, which had maintained a steady interest in FPC production since before the mid-fifties. Although the company had conducted laboratory process experiments and had taken out a number of protective patents, the Director of Technical Applications, H. M. Burgess, commented in July 1970 that interest was "rather minimal at the moment . . . It was one of those things that look awfully good on the surface, but where limitations have emerged that were not previously apparent."[126] Much of the private sector interest, including that of General Foods, had been discouraged by the failure of the production effort of Alpine Marine Protein Industries, Inc., under contract to meet their commitments to AID. Alpine had filed a complaint with the Armed Services Board of Contract Appeals regarding the termination of the contract with AID, making fourteen allegations against AID and asking for $5 million compensation. The matter was eventually settled out of court for a much lower amount, but not before VioBin had foreclosed on the plant at New Bedford, Massachusetts.

Many other companies (in addition to the VioBin Corporation) considered investing in some form of FPC development. Following are some of the more interesting examples:

1. *AVCO, Inc.,* hired one of the BCF scientists who had helped develop the IPA process to begin an FPC development program in 1967. By the end of 1968 the program had been abandoned with few tangible results.
2. *Cape Flattery Co., Inc.,* planned to build an IPA solvent extraction process aboard a 196-foot surplus Navy landing ship. The plant was to be capable of processing eight tons of fish per hour or 200 tons per day. A solvent extraction process was never installed, however, and so the ship only produced fish meal. In 1970, the company declared bankruptcy.[127]
3. *Food Protein Concentrates, Inc.,* a Culver City, California, concern specializing in producing fish meal from anchovy and hake, became interested in 1969 in designing a 3000 pound per day human-grade FPC to be produced for a market in Asia. There is no evidence that such a product was ever produced.[128]
4. *Oceanic Development Corporation,* a firm from Houston, Texas, planned to construct a 200 ton per day FPC plant in Baja, California during 1969. The plant was to employ the licensed VioBin process using anchovies as the raw material. It was never constructed.[129]
5. *ARA Services, Inc.,* a major institutional feeding company, proposed a demonstration school lunch program in Jamaica using FPC to be carried out by its subsidiary, World Food Systems, Inc. The project was, in fact, implemented in 1971 although lengthy start-up delays (due to political changes in Jamaica) pushed the start of meal production back to 1973. By 1974, 2500 metric tons of FPC (produced by the Astra Group) had been utilized in pastry dough as a replacement for soy flour. Although the feeding project is continuing, the use of FPC was terminated when its costs were found to be uncompetitive with soy flour as a protein source.[130]
6. *Nabisco–Astra* was formed when the Nabisco company had entered into an agreement with the Astra Group of Sweden in 1970 to produce products incorporating Astra's FPC for market testing and development. Since Astra was, as of 1975, the only major world industrial concern producing FPC, it was the hope of this company to combine the product development technology of Nabisco with its own FPC production skill.[131]

On the whole, it becomes apparent that private investment in FPC has been almost totally fruitless. Until 1973, the continuation of the packaging restriction set by the FDA was the principal reason for the inability of industry to develop a domestic market. As time passed without relaxation of the regulations, other issues also began to contribute to the negative investor attitude. Principally, these factors were economic in nature, namely, the cost of fish skyrocketed, the Japanese and Soviets moved in on available U.S. fish resources, processing costs increased due to the rising cost of energy, and labor costs also increased.[132]

In July 1971, however, the major obstacle remained the FDA ruling, still based, at least in part, on the agency's continuing suspicion about filth and the need to protect the consumer from it. The available evidence does not support the fact that FDA action on the removal of the packaging restriction was purposely withheld until all commercial interest had been eliminated. On the other hand, there is no doubt that the existence of the restriction after the 1967 ruling (which had approved the product as a food additive) obviously did serve as a powerful economic disincentive for commercial investment.

With the Alpine Geophysical Associates (the parent company of Alpine Marine Protein Industries) petition for the removal of the packaging limitation officially withdrawn on September 30, 1971, there was virtually no pressure on the FDA to modify its position. But at the end of December 1971, the fish meal industry group, the National Fish Meal and Oil Association, was persuaded by the NMFS to make one more attempt to create a domestic market for FPC by submitting another petition to the FDA for the removal of the packaging restriction.

EDP Shutdown

During the final stages of the initiative to gain the removal of packaging restrictions, the EDP operation in Aberdeen had actually been producing FPC. After reaching baseline operating conditions on June 15, 1971, some 1500 tons of Pacific hake were processed through September producing over 90,000 pounds of FPC meeting FDA requirements. The plant was plagued with raw material shortages and frequent mechanical breakdowns, and, as a result, the rate of production rarely approached the plant capacity of 50 tons per day. In fact, a subsequent Government Accounting Office report estimated that during the period from June to August 1971 it cost fully $5.08 for every pound of FPC produced, exclusive of depreciation and contractor fees.[133] It should be pointed out, however, that these costs do not accurately reflect commercial production figures since the plant was undergoing design modification throughout this period. However, it is still significant that the cost per pound was greater, by more than a factor of ten, than the costs originally projected for the product of $.25 to $.35 per pound.

In September 1971, a decision was made to alter the EDP for the processing of the so-called fatty fish: menhaden, herring, or anchovy. Menhaden was subsequently processed during the fall of 1971, although procurement and delivery problems required a switch to anchovy in early December. Anchovy processing continued until the plant ceased operations for design modifications just prior to Christmas. When operations resumed in mid-January using menhaden again, it became necessary to "import" the raw material from the Gulf Coast. At the beginning of March further processing problems related to the use of menhaden necessitated a second switch to anchovy in order to determine if the same or similar difficulties would be encountered with that species. Since the yields from both types of fatty fish had been extremely low,[134] the EDP was shut down on March 16, 1972, to discontinue the fatty fish operation and to prepare for a special lean fish run which, it was anticipated, would provide a minimum of 30 additional tons of acceptable product for distribution to AID and for research purposes.

Thus the EDP again operated during April and May using Pacific hake. Although insufficent data were generated on either lean or fatty fish to demonstrate the economic or engineering feasibility of FPC production, the need to produce an acceptable product for research and testing was made particularly acute by the approaching expiration of the EDP enabling legislation, PL 89-701, on June 30, 1972. Despite efforts on the part of NMFS officials, as well as favorable publicity derived from an international conference on FPC convened at MIT in early June,[135] congressional hearings were not convened in time, and the Aberdeen plant was forced to terminate operations and furlough its employees.

PROGRAM TERMINATION (1972): THE FAILURE OF BUREAUCRATIC SUPPORT AND POLITICAL WILL

Congressional Skepticism

One indication of the low esteem in which the FPC program was held by the Congress in 1972 is the fact that none of its previous legislative advocates was inclined to convene hearings on measures to extend the life of the demonstration plant prior to the mandatory shutdown of the EDP on the first of July. With the facility standing idle and all of the employees laid off,[136] there was little to compel legislators to act swiftly in considering an extension of the act. In one sense, the woes of the EDP can be attributed to shortsighted congressional action in the first place. NMFS officials often complained that, after losing a year and a half (after passage of the original enabling legislation) due to funding deficiencies and other technical difficulties, the statutory limits on the legislation were never revised. They argued that, if the plant had ultimately been given the chance to make up for the time lost, commercial viability could have been established.

Critics of the NMFS, on the other hand, have claimed that it was the shortsighted, ineffective leadership of the old BCF that was chiefly responsible for the premature termination of the EDP testing program. The real issue, though, would seem to be why the NMFS could not muster the necessary congressional support to gain an extension of the program, which, after all, is a fairly common occurrence in Washington (for example, in DOD programs).

In the vacuum left by the expiration of statutory authority for the EDP (and while waiting for the initiation of congressional hearings scheduled for August 1972), the NMFS proceeded with work on a six-month assessment of fish resource availability and cost and the development of an engineering model for FPC production. In addition, a contract was signed on July 7 with the Professional Engineering Development Company (PEDCO) of Cincinnati, Ohio, for an engineering review study of the EDP operating data for 1971–1972 (see exhibit 3). The study was to specify modifications that would make the plant capable of producing an acceptable grade product, while running continuously (one hundred hours or more) at design conditions and feed rates.

Finally, on August 11, hearings on the extension of the EDP were convened by John Dingell, then chairman of the Subcommittee on Fisheries and Wildlife Conservation of the House Committee on Merchant Marine and Fisheries. It was apparent almost from the start that Chairman Dingell and the rest of the subcommittee would have to be convinced that such an extension was worth supporting. Dingell in particular stated: "I think putting the plant on the West Coast was silly to begin with, and I think you are recognizing the fact that you made a bad judgment there . . . I'm not looking forward with any enthusiasm at all, I must tell you, to handling this bill on the floor, and I have not yet made up my own mind that I am not going to fight it."[137]

Realistic Strategies

Despite the difficulty of promoting a program before a hostile committee, various officials of the National Marine Fisheries Service and the National

Exhibit 3
Operations of the EDP, March 1971–May 1972

	Hake	Menhaden	Anchovy	Total
Number of months in which processed	8	5	3	
Tons of fish used	2,280.4	406.7	184.0	2,871.1
Tons of FPC produced	166.8	30.2	11.9	208.9
Average yield (percent)	7.3	7.4	6.5	7.3
Range of monthly percentage yield	0 to 10.2	3.6 to 11.1	2.5 to 6.8	
Tons of FPC meeting FDA specifications	46.95	.35	—	47.3

Source: GAO Report, 1973, p. 8.

Exhibit 4
Remaining Goals and Budget Needs for EDP Program

I. Goals

1. Modify the plant to handle fatty fish and complete the engineering evaluation tests.
2. Complete the plant processing economic runs needed to relate the production data to the overall commercial potential.
3. Produce additional quantities of the product from fatty fish to permit the completion of product acceptability and nutrition studies.
4. Finalize the technology on lean fish and prepare the final reports.

II. Projected Budget Needs

	First Year	Second Year	Total
Plant Operation	$270,000	$713,000	$ 983,000
Fish Procurement	562,000	257,000	819,000
Total	832,000	970,000	1,802,000

Source: U.S. , Congress, House, Committee on Merchant Marine and Fisheries, *Hearings Before the House Subcommittee on Fisheries and Wildlife Conservation of the House Committee on Merchant Marine and Fisheries,* 92nd Cong., 1st Sess., 1972, p. 245.

Oceanographic and Atmospheric Administration testified as to the remaining goals and requested budget for the EDP program. They suggested three principal strategies that the Congress might adopt concerning the future of the EDP: (1) leave the plant closed and simply terminate the project; (2) approve the continuation of the program but move the plant to some location nearer the supply of fatty fish; or (3) continue the program, leaving the plant where it was and have the fish shipped in.[138]

The alternative of moving the plant from the West Coast had come under active consideration mainly due to two factors: the inadequacy of the Pacific hake resource and the approval by the FDA of the use of menhaden and herring in the production of FPC. The use of the lean hake fish had always been considered an interim programmatic step because of the wider availability of fatty species and the attractive possibility of reducing production costs by utilizing the oil by-products of the fatty fish. Unfortunately, however, the fishing industry for fatty species was located primarily on the East and Gulf Coasts so either the raw material would have to be shipped out to the plant, with all of the attendant transportation and storage problems, or the plant would have to be moved nearer to the source of the raw material at great additional expense.

The Subcommittee directed the NMFS officials to prepare an analysis of the costs involved in moving the plant. Contact by the service with the NFMOA and the Gulf States Marine Fisheries Commission led to the identification of three geographical areas and seven specific sites in Louisiana and Mississippi near the menhaden resource.[139] Although the modular nature of the EDP facility made it possible to dismantle and move the equipment, none of the seven sites identified contained buildings, such as warehouses, suitable for housing the processing, storage, and research facilities.

In addition, a factor analysis of the costs of the proposed move indicated that it would require between $800,000 and $900,000 to dismantle, ship, prepare the site for, and reconstruct the EDP. These figures were compared with the cost of shipping the fish from the Gulf ports to Aberdeen, Washington, estimated at $340,000 (based on $65 per ton for 5200 tons), and the cost of freezing, packing, and handling the shipped raw material, amounting to $262,000 (based on $50 per ton for 5200 tons). Together, the total cost of having the necessary fatty fish shipped to the plant was approximately $600,000, thus effecting an apparent $200,000 to $300,000 cost savings[140] by pursuing the strategy of leaving the plant in place at Aberdeen.[141]

The NMFS representatives mentioned one additional factor, namely, that of the time required for the proposed move. It was estimated that the entire moving process would consume roughly sixteen months. Thus, by the time the legislation was passed, the necessary new equipment ordered, and the plant reassembled, as much as two years might have elapsed; this was compared with an estimated restart time of nine months for the Aberdeen plant.

The subcommittee was presented with various testimonies, both favoring and opposing the proposed move. As might be expected, the NMFS, on the basis of their cost analysis, came out strongly in favor of retaining the plant in its present location. Their position was endorsed by the NFMOA and the west coast fishing interests. The Haynie Products Corp. of Mississippi, on the other hand, urged consideration of the Gulf Coast sites where the EDP might conceivably be established in a symbiotic relationship with an ongoing fish meal operation. In addition, at the request of Congressman Dingell, the GAO reviewed the comparative analysis conducted by the NFMS, while simultaneously conducting their own investigation. The GAO report found that, despite certain errors in estimation, the NMFS was essentially correct in its calculation of the costs of moving the plant versus the costs of shipping the fish.[142]

It is somewhat curious that the NMFS had actually proposed moving the plant at all as a strategy for congressional consideration in view of the fact that: (1) it had not fully investigated the possibility, and (2) it had not made an *internal* policy decision regarding its own agency position. The Service clearly had not thought through the likely ramifications of its proposal, particularly the fact that it would place the agency in an untenable bind, namely, if it advocated moving the plant to the Gulf Coast it stood to lose whatever remained of the long and powerful support of Magnuson and Hansen (representatives from the State of Washington); but if, as eventually happened, it recommended retaining the present location, it stood to lose much potential new support from both the Gulf Coast congressional delegation and from private industrial interests. The very fact that a proposal had been made to move the plant was viewed by many in Congress as an admission on the part of the NMFS that the program was a failure. Finally, doubts about the significant time delay and actual resource availability involved in moving the operation may have influenced the final decision.[143]

FDA Intransigence

Other issues continued to concern the subcommittee about the FPC program. Once again the FDA's one-pound package restriction became the center of controversy. NMFS representatives testified as to the damage the ruling had inflicted on commercial hopes for the program; and Dr. Virgil Wodicka, Director of the Bureau of Foods at the FDA, stated frankly that "the intent was to make the product available directly to the consumer for formulations and incorporations in the kitchen, but not available to the industrial formulator."[144] This statement provoked intense questioning by Chairman Dingell concerning the motivation for the restriction. Pressing the FDA official closely, he demanded: "Where are you authorized to engage in that kind of harrassment, and what kind of administration is there that will do that? . . . It just strikes me, as far as the Food and Drug Administration is concerned, that this is the kind of small-minded administration that is present, and reflects no credit on your agency at all."[145] Dingell then ordered the commissioner of the FDA and the secretary of HEW to provide him with a complete written report on the facts behind what he saw as "plain and outrageous harassment of the industry by the one-pound package limitation."[146]

In a follow-up letter to the FDA commissioner, Dingell expressed his opposition to having the food and drug laws tampered with by Congress, but he also demanded swift and complete accountability for the manner in which the Food, Drug and Cosmetic Act had been applied.[147] The FDA, in a reply of September 26, 1972, agreed that the restrictions imposed on the product were more or less unprecedented, although some parallels could be found in the requirements for the packaging and distribution of margarine.[148] They indicated that the pending NFMOA petition for removal of the restriction was then under active consideration and would be acted upon at the "earliest possible date." It is likely that this intense congressional pressure did help to speed final FDA action on the packaging issue although another ten months passed before the ruling was finally promulgated.

Continued NMFS Efforts

The NMFS continued to make determined efforts to salvage the program by providing a variety of evidence including data on distribution of FPC produced at the EDP (see exhibit 5), and documentation of a recent attempt to incorporate FPC into a preschool snack program on American Samoa (chap-

Exhibit 5
Distribution of NMFS-Produced FPC as of December 11, 1972

The demonstration plant produced 47 tons of FPC meeting FDA standards. Eighty requests for samples received by the NMFS were filled as follows:

AID	62,000 pounds
Governor of Samoa	1,000 pounds
2 Government laboratories	550 pounds
46 industries, universities, and other organizations*	949 pounds
	64,699 pounds

Source: GAO report, 1973, p. 10.
*Total distribution to universities and private industry amounted finally to some 20 tons.

ter 10). The NMFS also noted the continuing lack of data to substantiate the commercial viability of the product. It had become evident, however, by the conclusion of the hearings, that the program was suffering from a terminal case of too many production problems and too little political support. The production problems at the EDP were well summarized in the report of the GAO: (1) plant and equipment design deficiencies caused severe sanitary problems and consequent product rejection; (2) poorly designed and selected equipment caused frequent breakdowns; (3) fish storage and handling methods caused product protein loss; and (4) production process was inconsistent.[149]

AID Program Failure

Yet another important factor in the subcommittee's final disposition of the FPC program was the testimony of the Director of Office of Nutrition for AID, Dr. Martin Forman. He reviewed the history of AID's involvement with FPC including: the sixteen country feasibility studies, the use/evaluation studies in Chile and Korea, the problems in procuring FPC through Alpine Marine Protein Industries, Inc., and the current, reduced status of the AID program. He reminded the subcommittee that Section 218 of the Foreign Assistance Act of 1968 encouraged the expenditure of some $2.5 million on FPC and other novel protein sources.[150] But, in spite of this testimony, the fact that the AID program had been prematurely and unsuccessfully terminated, after the overseas feeding potential and humanitarian qualities of FPC had been used as the principal rationale for the program for so many years, seemed to be a telling factor in the final decision of the subcommittee.

Final Decision

The political problems of the program were somewhat more obscure, although it is clear, in retrospect, that the coalition that had engineered the successful 1966 and 1968 congressional actions no longer existed. In the 1966–1968 hearings powerful influence had been exerted by Senators Bartlett and Magnuson on the Senate side and by Representatives Julia Butler Hansen and Hastings Keith in the House. In addition, Congressman Dingell, then the chairman of the key House committee, had not offered any serious opposition on either occasion, although he did express a growing degree of skepticism. By contrast, in 1972, the Magnuson-Hansen Washington state coalition had disintegrated with Congresswoman Hansen's announcement that she would not stand for reelection.[151] The final blow was the open opposition to the program in 1972 by Chairman Dingell, opposition based on the failure of the program to achieve its goals, the continued lack of relaxation in FDA packaging restrictions, the problems with the fisheries resource, and, ultimately, his own unwillingness to become the "laughing stock" of the Congress by bringing such a bill to the floor for the third time.

The FPC program had, quite obviously, gained the unenviable reputation of a political and technological white elephant. Few in the Congress, even from states where fishing was still a major industry, were willing to risk their reputations on a program with so dubious a performance record. With the old Magnuson-Hansen coalition no longer in existence and with any possible Southern support alienated by the position of the NMFS on moving the plant, there was just not enough interest or enthusiasm left in the Congress to see the program through to any sort of reasonable conclusion.

Thus, the 1972 House hearings did not result in a favorable report by the Subcommittee on Fisheries and Wildlife Conservation, and the Experiment and Demonstration Plant was not refunded. The facility at Aberdeen sat idle while certain tangential aspects of FPC research, already in progress, were carried on to completion in early 1973 at the College Park Laboratory in Maryland. Due to the need for major electrical and design modifications (estimated to cost up to $1 million), it was not even possible to sell the EDP to a private commercial interest. Eventually the facility, which had cost some $2 million to build and many times that to develop and operate, was sold for scrap to a San Francisco concern for a mere $255,000.[152]

EPILOGUE (1972–1975)

FDA Capitulation

The termination of the EDP program essentially marked the end of the U.S. Government's FPC development program. It is ironic, although somehow fitting, that just as the program was being phased out by the NMFS, the FDA was preparing to remove the bulk packaging restriction that had, for so long, served as an impediment to FPC development. The NFMOA had agreed in December 1971 to make one last effort to gain approval for the use of FPC in manufactured foods. Accordingly, a petition was submitted early in 1972 that requested that the FDA amend the Code of Federal Regulations to permit the manufacture, shipment, and use of FPC under commercial conditions. Unlike the earlier petitions, however, new evidence was presented that a significant percentage of the U.S. population was deficient in several nutrients, including protein. The NFMOA petition stated that a number of new foods, such as textured vegetable protein or single-cell protein, were being developed to meet this deficit and that FPC should be allowed to compete equally on its own merits.[153]

After the normal delays to fulfill FDA demands for additional technical information, the issue was finally resolved on July 17, 1973, with the publication of an order by the FDA officially removing the packaging restrictions from the commercial use of FPC.[154] The order stated that "these restrictions are no longer deemed necessary provided that the labeling of the manufactured foods containing the additive bears the declaration 'whole fish protein concentrate,' in the proper sequence in the statement of the ingredients."[155] The rationale given for the decision was that the name, "whole fish protein concentrate," had become so well known that it would no longer be deceptive to consumers, and thus, the need for packaging restrictions was removed. Thus, the FDA had ended up using one of its original arguments, prevention of consumer deception, to justify the complete reversal of its own policy. In fact, by June 1974 the FDA was itself advocating the use of FPC as a protein fortificant:

> The Commissioner recognizes that due to the rapid changes in food technology more protein sources will become available and that, therefore, it would be in the best interest of the consumer, from both nutritional and functional standpoints, to allow the use of other plant protein sources as long as it can be demonstrated that the materials are safe and suitable protein sources. In addition, animal proteins such as milk protein, *fish protein concentrate,* and egg protein can be used in the fabrication of a protein product and should be permitted.[156]

Private Sector Initiatives

There has been a limited amount of activity in the private sector since 1972 although it has yet to show significant results. However, two initiatives in particular are noteworthy. First, the Cereal Refining Company, Inc., of Portland, Oregon, developed a process in 1973 for chemically peeling wheat and other cereal grains, including rice, to which FPC could then be added. The venture utilized EFP-90 (FPC made from eviscerated fish) prepared by the Astra Group of Sweden. Due to lack of capitalization and market, however, the company has ceased operations. Second, the NABISCO-Astra Corporation now stands as the only significant ongoing attempt in the United States to develop a market for FPC.

The lack of a viable commercial market has, with the single exception of the Nabisco-Astra Corporation, been the undoing of every FPC development effort, including those directed primarily at humanitarian feeding programs. In its postmortem report on the U.S. development program, the General Accounting Office suggested that the market potential for the product had been limited by (1) the continued preference of food processors for functional ingredients (for example, soy concentrate) over the nonfunctional FPC produced at Aberdeen; (2) the fact that consumer protein demand was being met by other conventional and nonconventional sources; and (3) the impact of the omnipresent, one-pound package restriction.[157] These conclusions seemed to substantiate the earlier economic study conducted by Devanney in 1970 at MIT in which he found that FPC was not competitive with other protein sources.[158]

Conclusion

The GAO report further suggested that the ultimate justification for the FPC program as a boon to the flagging U.S. fishing industry had never been substantiated. In addition to the lack of lean fish, the harvest of the more abundant fatty fish, such as the menhaden, had reached maximum sustainable yield as a source of food, meaning that stocks used for other purposes (namely, fish meal) would have to be diverted for the manufacture of FPC. Further, there was never any demonstrated interest on the part of the U.S. fishing industry itself, only potential interest. In the limited experience that was accumulated in obtaining fish supplies for the EDP, it was found that U.S. fishermen had difficulty in converting from cost-conscious harvesting (based on which species brought the highest price) to volume-conscious harvesting (based on the total amount of a particular species caught). Particularly with regard to volume-conscious fishing, the minimizing of turnaround time (time spent in port unloading, refueling, and so on) is the key to success.[159] Thus, the fundamental economic criterion for the success of FPC was, and still is, the need to keep down the cost of the raw material. Unless current federal assistance efforts, such as Project Seafreeze,[160] come to fruition, the outlook for profitable FPC production in the United States must be considered very bleak indeed.

In examining the FPC program itself, it is important to keep in mind the scale of effort that was involved. During the entire period from 1961 to 1972 the total direct federal expenditure on FPC development amounted to approximately $14 million. (A budget summary of the BCF/NMFS program is presented as exhibit 6.) While this is certainly not a trifling sum, it is still quite insignificant when compared with the vast sums spent in other areas of the federal budget (namely, defense, space, health research). Thus, particularly from the point of view of Congress or the Office of Management and Budget,

Exhibit 6
FPC Budget Summary BCF/NMFS

Fiscal Year	Operation and Maintenance and Research	Reprogram and Supplemental	Subtotal	EDP Construction	Total
'62	$ 50,000		$ 50,000		$ 50,000
'63	120,000	$ 400,000	520,000		520,000
'64	446,000		446,000		446,000
'65	465,000	120,000	585,000		585,000
'66	465,000	66,000	531,000		531,000
'67	429,000	260,000	689,000		689,000
'68	429,000	700,000	1,129,000	$1,000,000	2,129,000
'69	1,135,000	720,000	1,855,000		1,855,000
'70	1,860,000		1,860,000	1,043,000	2,903,000
'71	2,138,000		2,138,000		2,138,000
'72	1,600,000		1,600,000		1,600,000
'73	435,000*		435,000		435,000
Total	$9,572,000	$2,266,000	$11,838,000	$2,043,000	$13,881,000

Sources: (1) Fiscal Years 1962–1970: Northrop Services, Inc., "Fish Protein Concentrate —Program Review and Plan," Part I, prepared for NMFS under contract No. 2-35399 by Northrop Services, Inc., Arlington, Va., pp. 2-6. (2) Personal communication from NMFS, 1976.
*These funds represented program phase-out costs.

this program was simply not important enough to justify (nor did it receive) concerted and continuous staff attention. It was supported and sustained (or, perhaps, overlooked) as long as it remained a political asset but promptly terminated as soon as it became a liability.

Except for the period during the Kennedy administration when the FPC program had a great deal of visibility, there is little evidence that upper-level managers within the bureaucracy ever took the time to examine the actual status of the effort to develop FPC. Optimistic status reports and future projections were accepted from the lower levels of the bureaucracy with little or no hard analysis of stated goals versus actual accomplishments. This situation obviously worked to the direct advantage of the lower echelon people who were principally concerned with maintaining the support and the autonomy of their respective agencies. Because the objectives and competence of the line agencies involved with the FPC program tended to vary greatly, the normally difficult task of interagency cooperation was rendered virtually impossible.

Certainly the performance record of the BCF indicates a definite lack of political, economic, legal, and even engineering capabilities necessary to deal effectively with the myriad of political, economic, and engineering issues involved in the development of a highly sophisticated and complex technology such as FPC. The organization of the Food from the Sea Service within AID reveals similar deficiencies, both in its overall size and in its specific expertise. The FDA, on the other hand, while well known for its careful and cautious analytic capability, often showed little or no sensitivity to the political and social issues involved in the FPC case. In addition, the particular problems that arose among the agencies (namely, the BCF, FDA, and AID) were characterized by the instrumental roles of individual personalities insofar as they were responsible for injecting personal prejudice and/or private ambition into agency operation and policies.

At almost every stage of the program the middle and lower levels of the bureaucracy within the various agencies were acting on distinctly separate agendas, which often were not entirely compatible with each other. One result of these conflicting purposes and the deficient state of agency communications was the unrealistically enthusiastic advocacy of Vice President Hubert Humphrey and the MSC for the FPC program. The council staff had inaccurately assessed the real capacities and commitment of the agencies involved in the Food from the Sea Program (due, in part, to misleading impressions conveyed by these agencies to the MSC), and, as a result, the MSC staff did not foresee the likely effect of implementing a premature interagency arrangement. Consequently, the Food from the Sea Program soon proved both ineffective and unmanageable.

But the most serious outcome of the bureaucratic failure was the inability of FPC proponents to mobilize, expand, and maintain political support at crucial points of program evolution. FPC, as a predominantly "social" technology (as distinguished from technologies related primarily to the defense or space sectors), never enjoyed the benefits of "the convergence of technological opportunity and national need"[161] commonly found in other areas of technological innovation in the postwar period. Instead, advocates were compelled to wage a never-ending effort to maintain an institutional base of support for the program. Yet, by and large, these individuals were neither qualified nor capable of entering the political arena in an appropriate and effective fashion; and, in fact, a clearly defined bureaucratic strategy for the promotion of the program was rarely, if ever, apparent. It has been suggested that "politics is a systemic requirement. What distinguishes programs in government is not that some play politics and others do not, but, rather, that some are better at it than others."[162] Ultimately, the inability of FPC program managers to generate and maintain a substantial public and private sector constituency caused the project to be terminated while other more expensive and technically less successful programs were, and are, continued.

Finally, a backward glance at the record of the U.S. FPC program indicates that it did *not* ultimately achieve its stated objective of demonstrating the commercial feasibility of FPC, due to high cost and relatively low yields of the product at FDA-approved levels. It may be argued that this result was inevitable because of the problems encountered with the EDP, such as the claims that the plant was designed and constructed on the basis of incomplete, and in some cases incorrect, engineering data. No doubt, the poor design of the EDP facility, which resulted primarily from the lack of in-house engineering experience, added greatly to the cost of construction, due to the need to employ expensive, outside consulting assistance and the necessity of costly, experimental processing equipment.

The engineering failure of the EDP is indicative, however, of a larger problem that was evident throughout the lifetime of the public sector FPC program. At the BCF (later the NMFS), at AID, and even at the Marine Sciences Council there was a fundamental misapprehension of the political and bureaucratic realities involved in the successful implementation of a technological innovation such as FPC. Because so little capacity or concern for the effective management of the uncertainty inherent in the program existed within these agencies, opportunities to solidify political support were missed, serious tactical errors in program operation were committed, and program policy was formulated in a disjointed, incremental fashion.

The FPC program failed ultimately to demonstrate many of the characteristics essential to the successful commercialization of a new technology: mastery of technological obstacles, cost-and-risk sharing with private sector participants, project initiative at the local level, a strong industrial system for commercialization, participation in the demonstration by those who would take responsibility for further diffusion, and the absence of tight time constraints. In fact, the obverse of many of these characteristics existed in this case in unfortunate abundance.[163] Perhaps these problems can be attributed, in part, to the technological optimism of the period in which the FPC program came into existence, or to the premature politicization of the program, or to the lack of sufficiently qualified people in key positions capable of asking the really "hard" questions regarding program design. In all likelihood, however, it was a combination of all these factors that caused a program begun with such great expectations to be terminated with such disappointing results.

Chapter 4
FPC and Bureaucratic Decision Making: Promoter Versus Regulator

The arduous twelve-year (1961–1973) interagency struggle between the U.S. Food and Drug Administration (FDA) and the U.S. Bureau of Commercial Fisheries (BCF), later the National Marine Fisheries Service (NMFS), over the approval for commercial use of whole fish protein concentrate represents an important and illustrative example of many of the characteristics of bureaucracy that often serve to impede social and/or technological "progress." An examination of these characteristics, from the perspective of both the promoter (BCF/NMFS) and the regulator (FDA), is worthy of further attention here in the context of how such problems might relate to other nutrition innovations pursued through similar channels. By examining the principal motivations and pressures (as well as the general political environment) affecting both of the actors at each stage of the program's development, it will be possible to isolate the crucial points at which the program was either recklessly promoted or unnecessarily restrained.

A logical point of entry to these matters is a consideration of the question often posed by critics of the FPC program; namely, if the original justification for the product was to alleviate protein malnutrition in the developing world, why was the attempt ever made to seek the approval of the FDA when the regulatory agency must, de jure, make its rulings according to standards of acceptability for use in the United States? The most simplistic response to this query is to suggest that the original developers of the product did not consider, perhaps naively, that the acceptance of FPC—a product derived from a ubiquitous food source with diverse cultural connotations—would prove to be so controversial an issue for the U.S. government and public at large.

A more critical analysis of the reasons that the FDA was initially approached reveals a number of additional factors. For one thing, whether the BCF ever intended to get involved with the FDA concerning the approval of whole FPC became largely academic once it became known that the VioBin Corporation already planned to submit a proposal to establish a food standard of identity for the product. The history of prior conflict between the president of the VioBin Corporation, Ezra Levin, and the FDA was of substantial concern to BCF officials, who believed, with a certain degree of justification, that, if Levin planned to petition on behalf of FPC, the BCF must also become involved in order to ensure that the product was not rejected solely on the basis of reputational issues related to VioBin.

Another motivation for direct BCF involvement in the FDA petition derived from the fact that the BCF, as a government agency, could not (nor did it wish to) produce FPC commercially. There was, therefore, an obvious need for the participation of private industrial interests to insure production and distribution once the product had been sufficiently developed. However, substantial commercial involvement of a U.S. food processing (or engineering) company with a product not approved by the FDA was extremely risky and virtually unprecedented. This state of affairs provided another powerful incentive for seeking FDA approval.

In terms of the humanitarian objective of food aid suggested earlier, it is generally well known that the majority of world standards for food purity and acceptability are modeled (often unnecessarily) on the criteria developed by the U.S. Food and Drug Administration. Once it became apparent to the BCF that FPC would have to be presented for consideration to the FDA, the international influence of that agency's standards became extremely significant. In the view of the Bureau it seemed highly unlikely that a food product would be accepted for use in a developing country, even for disaster or famine relief, if it were judged unacceptable by the FDA. Thus, once the subject had been broached, it became imperative to convince the regulatory agency to approve the product.

One final consideration which, though more or less unstated, also figured in the BCF decision matrix was the growing awareness of the scope of potential U.S. domestic demand. Although popular thinking in 1961 still clung to the notion that little (if any) protein malnutrition existed in America, many individuals, both within and outside the government, felt that a possible need existed for a low-cost protein supplement to be used in institutional feeding and/or direct sale to low-income families. Obviously, for such a product either to be marketed by a private company or to be distributed by the U.S. Department of Agriculture (USDA), formal FDA approval would be required.

INITIAL FDA REACTION

As we have indicated, a decision by the BCF as to the advisability of petitioning the FDA for approval of whole fish protein concentrate was largely preempted by the filing of a petition on behalf of the VioBin Corporation in July 1961. The reaction of the FDA to this first initiative was an informal ruling that the product violated Section 402(a)(3) of the Federal Food, Drug and Cosmetic Act of 1934 because it contained parts of the fish that were considered filthy, and thus the product itself had to be considered adulterated. Despite the fact that, according to this ruling, FPC could not be sold in interstate commerce, the administration indicated that it would not interfere with its sale overseas under the terms of Section 801(d) of the Act.

The FDA is not insensitive to the impact of its rulings on the regulation of food products in other nations. It is quite likely, therefore, that the administration realized the improbability of a product succeeding in an overseas market when it was explicitly banned in the United States. However, the agency did attempt—one must assume in good faith— to suggest some alternate courses for the FPC proponents. They might (1) seek the specific congressional amendment of Section 402(a)(3) of the Food, Drug and Cosmetic Act so that whole fish would no longer be considered adulterated; (2) seek the resolution of the issue through the courts; or (3) seek administrative remedy through the FDA in the form of a definition and standard of identity for FPC.

It was at this point that the Bureau of Commercial Fisheries formally became involved. They had rejected the FDA's first two recommendations as infeasible and had opted instead for the third by publishing a proposal, on September 15, 1961, to establish a definition and standard of identity for a product called "fish protein concentrate, whole fish flour."

It is by no means clear that this initial BCF policy decision was a wise one, especially since it was apparently made without the benefit of competent legal counsel. But, by choosing not to test the issue in court,[1] the Bureau, in effect, played directly into the hands of the FDA, allowing *them* to call the shots. In fairness, however, it must be added that, without the existence of what the Bureau considered to be an acceptable product (the VioBin FPC was already available but did not meet BCF standards), there was little likelihood of a test marketing.

It is also evident, in retrospect, that the BCF did not aid its own cause in the name chosen for the food standard proposal, "fish protein concentrate, whole fish flour." During the sixty-day comment period which followed the publication of the proposal a significant number of the adverse responses received by the FDA represented elements of parallel or related industries—the baking, milling, and dairy industries—which considered that the name "whole fish flour" would impugn the high standards of purity maintained for grain and dairy products. This type of industry reaction gave FDA officials, who were already on record as considering the product adulterated, the kind of supporting evidence that they had been seeking. FDA Commissioner George Larrick (who had begun his own career with the FDA as a seafood inspector) cited as justification both the formal objections and the alleged violation of Section 402(a), clause 3, of the Food, Drug and Cosmetic Act upon issuing a modified definition and standard of identity on January 25, 1962, that did *not* permit the use of whole fish.

According to the ruling published in the *Federal Register*[2] so-called "fish flour" could be made only from edible species of *eviscerated* fish after discarding the heads, tails, fins, viscera, and intestinal contents. The *Federal Register* listed four principal statements concerning the revised standard:

1. Consumers would regard the product as filthy,[3] and thus it would be in conflict with Section 402 (a)(3) of the Food, Drug and Cosmetic Act.
2. It would not "promote honest and fair dealing in the interest of consumers" to establish a standard of identity for a product regarded as filthy by most Americans.
3. Since Section 801 (d) of the Food, Drug and Cosmetic Act permits the export of a product that does not meet U.S. standards (as long as it does not violate the laws of the importing country), FPC could be used to meet the protein needs in less developed countries.
4. Even though there was "no evidence" of protein deficiency in the United States, there was still a reasonable basis for establishing a standard of identity.[4]

The FDA response represented an astute political strategy; namely, while ruling out the use of *whole* FPC (the only form in which the product was economically feasible), the agency continued to maintain the appearance of objectivity by sanctioning the export of the product and the acceptance of an eviscerated, domestic version. It is significant that even the agency's principal objections to whole FPC did not, at least initially, involve empirical considerations; rather, they concerned aesthetic judgments of what the consumer would *regard* as filthy, as well as vague, subjective notions of honesty and fair dealing "in the interest of consumers."[5] The FDA thus had won the first round of what FPC advocates in the BCF were beginning to realize might well be a long and costly battle.

ENDORSEMENT FROM THE NATIONAL ACADEMY OF SCIENCES

The Bureau was not alone in its unhappiness over the FDA response, and formal objections to the modified food standard proposal were filed shortly thereafter by a variety of industrial concerns and trade groups. As a result of these objections, the FDA stayed its order on FPC and scheduled hearings on the matter for June 18, 1962. This action prompted a committee of the National Fish Meal and Oil Association (NFMOA), an industry trade association, to convene a series of urgent meetings with representatives of the BCF to prepare testimony for the hearings.

The hearings were never held, however, due to the intervention of Interior Secretary Stewart Udall, who requested the National Academy of Sciences–National Research Council (NAS) to appoint a fact-finding committee to study the issue. The FDA responded to the Interior Secretary's initiative by ordering the hearings stayed indefinitely pending the outcome of the NAS review.

While survey and preliminary process research moved forward at the BCF College Park Laboratory facility, public debate over the viability of FPC continued, in anticipation of the NAS report. These arguments fell principally in four substantive areas:

1. Aesthetics: the FDA claimed that the inclusion of heads, tails, fins, viscera, and intestinal contents would result in a finished product that was filthy or otherwise unfit for food. Proponents countered that the finished product could not be identified as having come from any particular part of the fish and therefore should not be considered objectionable.
2. Regulatory Inconsistencies: Proponents claimed that other foods made from or composed of questionable materials, such as gelatin, sausage, clams, or oysters, were not considered aesthetically objectionable and that, in any case, proper labeling of FPC could meet this objection. The FDA responded that such foods were approved on "historical grounds" (long use) since consumers knew what they were eating. Furthermore, the FDA stated that even with proper labeling consumers might be "fooled," since they might unknowingly eat the product either in manufactured foods or in a restaurant.
3. Urgency of Approval: the FDA questioned why a standard was being sought when no suitable product was available nor a satisfactory commercial process in hand. In addition, it claimed that the urgency of the need did not justify relaxation of food standards. Proponents countered that private industry would not risk venture capital without FDA approval. They pointed to the deteriorating state of the U.S. fishing industry, which potentially stood to benefit from FPC production.
4. Need and Use Outside the United States: The FDA saw little need for FPC as a protein source in the United States, and they suggested that the product be sold only overseas where there was a clear need. Proponents replied that to export a product without FDA approval would provide an appealing target for communist propaganda. By the same token, shipping a filleted FPC product meeting FDA requirements would so increase the price as to make it prohibitively expensive for use in a less developed country (LDC).

In October 1962, the ad hoc NAS committee reported that a safe, wholesome, and nutritious whole FPC product could be produced. On this basis, it recommended that the FDA hearings be stayed indefinitely until a suitable extraction process was developed and the product adequately tested.

Partially in response to the principal recommendation of the NAS ad hoc committee the Kennedy administration recommended a major increase in the FY '64 BCF appropriation earmarked for FPC. After receiving congressional approval, the Bureau was able to utilize the additional $400,000 to commence a full-scale laboratory development program on FPC. The results of this work were presented to the NAS committee at its meeting on September 16/17, 1965; on the basis of this new evidence, the scientific review board was able to state categorically that the FPC developed by the BCF was "wholesome, highly nutritious, stable, and safe" and that it was not "in any way filthy or otherwise unfit for human consumption."[6]

FPC: FOOD OR FOOD ADDITIVE?

Despite the significant new data amassed by the BCF in the course of its most recent research, the FDA position remained unchanged even after the agency had had the opportunity to review the BCF findings. A memorandum, containing the preliminary reports of the various FDA divisions, exemplified the agency's position: the Division of Toxicological Evaluation voiced no objection to the conclusion of minimal health hazard in FPC; the Division of Nutrition found FPC equal to casein, although data was lacking on the effects of intestinal bacteria; the Division of Food Standards and Additives, however, continued to report that FPC violated Section 402 (a)(3) of the Food, Drug, and Cosmetic Act, thus causing the product to be considered adulterated, or in the words of the FDA, "a crime against decency."[7]

Pressure for resolution of the FDA impasse began to mount with the formal presentation of the NAS committee's report concerning FPC to Secretary Udall on December 1, 1965. A high-level BCF policy meeting was convened on December 3 to determine the correct response to the FDA, in light of White House pressure to release the NAS report to the public. It was decided that BCF Director Donald McKernan would contact FDA Commissioner George Larrick to inform him of the situation. However, a subsequent telephone call revealed that George Larrick had left the FDA on terminal leave prior to his retirement. (The fact that the BCF leadership did not even know of Larrick's departure is indicative of the state of relations between the two agencies.) Larrick's acting replacement informed McKernan that three possibilities were then under active consideration: (1) FPC might be approved for sale but limited to well-labeled, individual packages and not allowed in a mixed product; (2) the Food, Drug and Cosmetic Act might be amended by Congress to exclude FPC; or (3) the FDA approval might be limited to one species and one product. There was also discussion of the possibility of approving whole FPC but subjecting it to clearance under the procedures of the

Food Additive Amendment.[8] A meeting with representatives of the FDA was scheduled for later in December.

It was obvious that, despite the departure of arch FPC foe George Larrick, the policy position that he had developed regarding FPC remained in effect. However, under mounting pressure from the BCF and the White House over the pending release of the NAS report, the regulatory agency could not expect to maintain its intransigent position indefinitely. It is within this framework that we see developing the three preliminary FDA strategies, designed to strictly limit the approval of whole FPC.

The long-awaited FDA–BCF meeting was finally convened on December 9, 1965. It began with the BCF outlining its problem with regard to the pressures being applied by Secretary Udall for early release of the NAS report. The FDA, in turn, reiterated the tentative conclusion of its scientific staff, namely, that the safety of the product was established but that it must still be considered adulterated. The administration emphasized that the ultimate decision on acceptability of FPC was the responsibility of the FDA alone, and that the recent endorsement of the NAS committee, while given much weight, would neither prejudice nor accelerate their decision. The BCF petition would, however, be given top priority action.[9] It would appear that the position of the FDA in this instance was justified in that the findings of the NAS committee were based not on new scientific research but on a review of work previously completed.

An FDA representative then revealed a major shift in policy of the regulatory agency: since FPC had not yet been marketed as a human food product, the FDA now believed that it should be considered a *food additive,* rather than a whole food, under the Food Additive Amendment, Section 201 (d) of the Food, Drug, and Cosmetic Act. The distinction between food and food additive is significant. Whereas a food, once accepted as meeting certain standards, is not subject to on-going FDA regulation, a food additive—which is not considered a food commodity in and of itself—is subject to rigorous and continuing regulation. Thus, if the FDA ultimately had to approve the product, it could in this way maintain strict surveillance over its manufacture.

It was suggested that there was ample precedent for another government agency petitioning for such a ruling. While, by law, the FDA must consider the objections of the consumer to a food additive petition, as a rule the agency did not do so if the objections were considered frivolous. The BCF representatives were assured that since the inception of the Food Additive Amendment, no public hearings had ever been necessary.[10]

The Food Additive Amendment had been enacted on September 6, 1958, and was implemented gradually, becoming fully operational in June 1962. This amendment completely altered the government's method of regulating the use of additives in foods. It provided, for the first time, that an additive could not be used in foods unless the FDA found that it was safe at its intended level of use. Exception was made for food additives which, because of years of widespread use, were "generally recognized as safe" (GRAS). These items did not have to undergo extensive testing and no specific tolerance levels were set for them, that is, no maximum allowable concentration was specified.

The FDA–BCF meeting terminated after discussion concerning the relationship between a food additive, under Section 201 (a) of the Food, Drug and Cosmetic Act, and the marketing restrictions that were likely to be placed on the product. BCF representatives were assured that, in the very near future, the FDA would develop formal positions on the questions of a food additive petition and of the status of FPC as an adulterated food. A follow-up meeting was subsequently held on December 28, at which time BCF officials were informed of the data that would have to be presented in support of a food additive petition, including data on IPA residue levels from the solvent extraction process.

The BCF leadership was thus faced with an extremely difficult decision, namely, whether to abandon the food standard of identity proposal in favor of a food additive petition. Besides the cost to the Bureau in terms of further implementation delays and valuable research time and money spent in responding to FDA information requests, there was also the overriding concern that FPC was a *food* (that is, something that could be consumed by itself), not a chemical additive, and must be considered as such. Before deciding which course to pursue, McKernan sought advice from a number of sources. He first convened a fishing industry meeting in January 1966, which provided verbal encouragement but little in the way of financial support or actual involvement. Since one federal agency obviously could not take another federal agency to court, only the private sector was in a position to challenge the FDA position. But industry has shown little inclination to become involved in litigation of this nature, primarily because it is costly, time consuming, and requires product marketing and seizure. Besides the negative publicity accruing from such an initiative, tenuous court challenges tend to find little favor with profit-motivated investors.

McKernan did finally seek competent legal advice regarding the advisability of pursuing the food additive petition from a well-known Washington food and drug lawyer, Vincent Kleinfeld. Kleinfeld suggested that the FDA was probably using the additive approach as a means of postponing action. If this were the case, it could leave the Bureau in an extremely vulnerable position since, even after examining the petition, the FDA might *still* turn it down by saying that FPC made from whole fish was adulterated. The food and drug lawyer indicated that he was impressed by the recent NAS report and that, on the basis of past FDA reliance on such NAS studies, this should be considered sufficient evidence of safety and wholesomeness.[11]

Kleinfeld urged that the BCF not submit a formal food additive petition until the FDA gave definite assurance, in writing, that it intended to approve whole FPC. He also recommended several additional strategies designed to force the issue with the FDA.

1. Secretary Udall should write to Commissioner Goddard (the new FDA commissioner) asking for confirmation that FPC made from whole fish would be approved.
2. Senators should be urged to write to the FDA demanding assurance that FPC would be approved if a food additive petition were presented.[12]
3. As a last resort, the White House could be requested to pressure Goddard into certifying FPC as acceptable.

Kleinfeld was especially wary of engaging in the petition procedure before receiving assurances that the FDA would approve whole FPC because, as he explained, objections raised to such a petition would require the FDA to hold hearings, which could delay approval by a number of years.[13]

Ultimately the BCF leadership found it impossible to follow Kleinfeld's advice. Subsequent events, including the so-called "War on Hunger" speech by President Lyndon Johnson and the public release of the NAS subcommittee report on FPC, had combined to significantly reduce the options available. Thus, despite the fact that many on the staff of the BCF still believed that FPC should be considered a *food* and not an *additive*, and without the benefit of *formal* legal counsel, the Bureau decided to abandon the effort to obtain a food standard of identity for FPC in favor of seeking its approval as a food additive.

THE TECHNICAL ISSUES

Once the decision had been made to proceed with the food additive petition, work at the Bureau's College Park Laboratory was immediately directed toward producing the kinds of additional supporting data required by the FDA for food additive petitions. After an FDA request for more complete data on lead residues was fulfilled,[14] another interagency meeting between FDA and BCF representatives was convened in April 1966. At that time the FDA expressed its concern about the significance of the lead content (7 ppm where 0.6 ppm was considered safe) as well as the isopropyl alcohol (IPA) levels (7000 ppm where 1400 ppm were considered safe) based on FPC solid level in the diet of 20 percent. For the first time the FDA also expressed concern about the fluoride content in FPC (up to 12 milligrams per day where 3 to 5 milligrams were considered safe), which was derived from the presence of bones and scale. Representatives of the regulatory agency stated that, if the IPA, lead, and fluoride levels could not be substantially reduced, chronic toxicity studies (involving two-year feeding tests) would become necessary. The FDA also announced that it was contracting with Albany

Medical College for studies on monkeys and humans to gain data on the metabolism of IPA and lead residues.

The BCF, for its part, questioned the validity of the FDA's assumption of a 20 percent FPC dietary solids level, claiming that use of the product above 10 percent was neither economically nor technologically viable. In response to the BCF criticism the FDA pointed out that, in certain malnourished areas of the developing world, high levels of FPC might be consumed by very hungry people who had no other source of protein. This sudden concern of the FDA about the international use of the product is strangely at odds even with the agency's own stated policy position from the same meeting.

> The Commissioner is aware of the importance of the work you people are doing to supply a nutritious product which is so necessary to many parts of the world. The FDA scientists are also aware of the importance of this work and we would not hesitate to recommend its use elsewhere (in the world) but we must be able to assure the Commissioner that it is safe and we must apply the same regulations to this as we do to other food additives. From your point of view also it would be a disaster if we could not say the product was safe.[15]

As a result of the FDA demands, the BCF commenced a new series of tests to determine the fluoride and lead levels, resulting from the IPA extraction method, and to ascertain whether the IPA residue could be substantially reduced. The new BCF data and revised specifications apparently relieved the principal FDA concerns regarding IPA and lead residues. However, based on the studies conducted by Dr. Isadore Zipkin of the National Institutes of Health,[16] the FDA advised the Bureau of the need to materially reduce the fluoride content in FPC. In a follow-up FDA–BCF meeting on July 25, 1966, FDA Commissioner Goddard officially reported that the levels of lead and IPA were no longer a problem, but that the agency now had new and serious concerns with the levels of fluoride in the FPC it had tested. The concerns did not involve fluoride toxicity, but rather the problem of cosmetic mottling of teeth.

It is unfortunate, in retrospect, that any analysis of the FDA position on this issue is clouded by the controversy that raged during this period concerning the addition of fluorine to public water supplies. Was the FDA simply covering itself in the event that the occasionally irrational antifluoridation forces might turn their wrath on the FDA for permitting a high-fluoride FPC product on the market? As with so many of the issues surrounding the FPC program, there is no definitive answer to be found in the existing evidence.

An Interior Department official complained to the FDA about these further delays and pointed out that the chances of U.S. families consuming more than 5 percent FPC regularly in their diets was remote. Commissioner Goddard responded that the FDA could not allow a product on the market that would cause mottling. The FDA commissioner assured the group that the fluoride problem was the last remaining obstacle to FDA approval and that if the fluoride content were reduced to 75 to 80 ppm (at 50 percent availability), the FDA would be able to approve the product. Four days later, on July 20, 1966, Undersecretary Carver informed the FDA that BCF could meet the 80 ppm standard for fluoride and that the FDA should proceed immediately to draw up regulations.

Many individuals involved at the time, as well as those who have examined the program since its demise, have questioned whether the FDA's introduction of the technical problems—with lead and fluoride content and IPA residues—was a legitimate issue or simply a delaying tactic. It was pointed out, for example, that, while the BCF maintained that FPC would not (and could not) be used at levels greater than 10 percent of dietary solids, the FDA continued to assume a dietary level of 20 percent, thereby exaggerating the potential role of FPC in the diet.[17] Some critics, such as the late fisheries expert Wilbert Chapman, have also claimed that the FDA was attempting to protect the dairy industry by its actions.[18]

The total evidence available does not appear to support the imputation of such ulterior motives to the FDA policy makers. While there is no doubt that a great deal of unnecessary foot dragging and obfuscation occurred regarding the approval of the product,[19] the BCF's own actions were often counterproductive. For example, the technical data released to the FDA was often overly rigid (and occasionally incomplete), having been produced from samples manufactured under sterile, laboratory conditions which could not be duplicated commercially.[20] This resulted in unrealistically low FDA standards which ultimately forced the addition of a large amount of capital-intensive equipment to the final demonstration plant, thereby driving up the cost of production and discouraging private investors.

THE BULK PACKAGING RESTRICTION

Final agreement had not been reached on the fluoride question when the FDA voiced yet another concern for the protection of the "unknowing consumer." The regulatory agency stated that it was considering the imposition of a bulk packaging restriction, that is, limiting sales to small, clearly labeled, consumer quantities of one-half or one-pound packages to effectively prevent the use of FPC in manufactured foods where the consumer would be unaware of its presence. The threat of yet another FDA obstacle was of great concern to the new BCF Director, Harold Crowther, who had succeeded Donald McKernan in September 1966.

Crowther and others at BCF held a series of informal meetings in November and December with food and drug legal experts concerning the legality of the FDA's proposed packaging restriction of FPC. The idea of the FDA approving a product as a food additive but attaching restrictions so that it could not be *added* to food (by the food industry) seemed, from the perspective of the BCF, to be incongruous to the point of absurdity! The consensus of the meetings was that the FDA would be in an extremely weak legal position on the packaging issue, especially if the product were clearly labeled.

The food lawyers noted that the FDA's claim that the one-pound limitation was necessary because consumers do not read labels was contradictory to the agency's historic insistence on accuracy and clarity of labels based on the reverse argument, namely, that consumers *do* read labels. "The FDA can't have it both ways," they argued.[21] In another meeting food and drug attorney Michael Markel suggested that the FDA had two choices: (1) reject the food additive petition, which would result in public hearings; or (2) develop and publish FPC regulations, which the FDA would then have to defend.

Markel stated that the limitations on bulk sales gave the food agency an out not only on the "whole fish" issue but also on the need for public hearings, since all effective arguments would be eliminated.[22] He recommended that, if at all possible, the BCF should concede on the one-pound package restriction, temporarily, with the thought that the issue might be more effectively raised again at a later date and settled "out of court," that is, without a hearing. Markel agreed with the other opinions that the FDA was in a weak legal position regarding its statutory authority, and he urged the Bureau to retain expert legal representation if it desired to press the issue. Finally, Markel indicated that if a satisfactory compromise could be arranged with the dairy industry concerning its expected demand for hearings on FPC, the need for such an expensive and damaging procedure would be eliminated.[23]

THE 1967 DECISION

All of the uncertainties and unresolved issues between the FDA and BCF were finally brought to a head by the FDA announcement on February 2, 1967, of the tentative approval of the BCF petition for the manufacture of whole FPC using fish of the hake family. In the process of sanctioning the use of FPC the regulatory body imposed several significant restrictions:

IPA levels were limited to 250 ppm

Ethyline dichloride (used in the VioBin process) was limited to 5 ppm

Fluoride levels were set at 100 ppm

Total FPC in the diets of children up to eight years old was not to exceed 20 grams per day

FPC could be sold directly to the consumer only in clearly labeled one-pound (or less) packages

In the preamble to the ruling, the FDA disclosed its rationale for the ban on bulk sales and consequent use in manufactured foods: "Individual consumers are entitled to the opportunity of choice in the matter of whether they desire to include this item in their diets. . . . Bulk sales or packaging in units larger than one pound cannot be authorized unless the Commissioner is supplied with data demonstrating that the proposed use of the article will promote honesty and fair dealing in the interest of the consumer."[24] Once again, a vestige of the earlier filth issue was influencing the thinking of the FDA. In the words of one official the one-pound package limitation was imposed "as a means of dealing with the substantial number of consumers who do not wish to eat a product which contains viscera and intestinal contents."[25] As suggested earlier, the FDA, by this ruling, was in effect approving the use of a food additive which, except in special circumstances, could not be added to foods on an industrial scale. Since the regulatory agency had at the same time certified the product as safe, a serious question was raised as to the legality of prohibiting the use of FPC in manufactured foods. No one was *forcing* the consumer to use the product. Thus, if it were clearly labeled, how could such constraints on use be legally justified? The obvious implication of the ruling was to suggest a more covert FDA motive, namely, to approve the use of the product but to place such severe restrictions on its use that all commercial involvement would be effectively thwarted.

The FDA order provided that anyone adversely affected could file an objection within thirty days. Over 125 such objections were received (see exhibits 1, 2, and 3), more than 90 percent of which were from the various national dairy interests. In fact, Charles Fistere, an attorney representing the American Dry Milk Institute, requested that the BCF prepare a one-to-two pound sample of FPC for chemical analysis preparatory to filing a formal objection with the FDA. This Chicago-based trade association feared that FPC would unfairly compete with the world market for dried milk since, in their view, it was not produced according to the same rigid standards used for milk. In June 1967, the National and Regional Associations of Food and Drug Officials also voted formal resolutions of objection to FPC as a food additive. (See exhibit 4.)

Counterpressure arose from Senator Warren Magnuson (Washington) and other interested parties for a swift decision as to whether the FDA would hold hearings on its February 2 FPC ruling. On July 11, the FDA confirmed that, despite the original intention of the agency *not* to hold hearings, it would be compelled to do so because of the vigorous objections to FPC raised by the American Dry Milk Institute. FDA staff counsel recommended that the following four issues be considered:

1. whether the name "whole fish protein concentrate" would promote deception of the consumer;

2. whether the incorporation of whole fish causes a food additive to consist of filthy substances and therefore to be adulterated under Section 402 (a)(3) of the Food, Drug and Cosmetic Act;
3. whether the regulation fails to provide adequate safeguards against excessive intake of fluorine and heavy metals; and
4. whether the proposed use of the additive would promote deception of the consumer in violation of the Food, Drug and Cosmetic Act.[26]

A serious threat to the FPC program again loomed, but deliverance this time came from Vice President Hubert H. Humphrey, who, as a former senator from the dairy state of Minnesota, was sensitive to the needs and concerns of the dairy interests. To convince the various dairy groups to withdraw their formal objections, he suggested that the uncoordinated, and often conflicting, standards for milk products promulgated by the USDA and the FDA might be favorably resolved. According to Edward Wenk, Jr., the dairy interests were sufficiently impressed with this proposal to forego their opposition to the approval of FPC.[27] Thus, the need for FDA hearings was eliminated when the principal objector, the American Dry Milk Institute, formally withdrew its protest citing, as an official justification, the pressing world food need that FPC might conceivably help to alleviate.[28] However, it was not until March 2, 1968, that the FDA issued a final decision that the objections to the FPC regulation were "without merit."

EFFORTS TO BROADEN THE SCOPE OF FDA APPROVAL

The BCF initiated meetings with the FDA in August of 1967 regarding the expansion of the approved species list from hake to include the use of the so-called fatty fish, such as menhaden or herring. During the spring of 1968, Alpine Marine Protein Industries, Inc., the firm contracting with AID to produce FPC for overseas feeding, also made inquiries as to the requirements for a petition to approve the use of menhaden and other fatty fish.

The fluoride issue continued to stir debate despite the 1967 FDA ruling setting the limit at 100 ppm.[29] A number of noted scientists in the field of nutrition wrote letters to BCF Director Crowther commenting on the unnecessarily restrictive FDA-approved fluoride level. For example, Dr. Nevin S. Scrimshaw, chairman of the Department of Nutrition and Food Science at MIT, wrote that not only would fluoride levels of 250 ppm be safe and nontoxic, but that the present 100 ppm regulation would only serve to drive up the price and decrease the effectiveness of FPC in preventing dental caries. Scrimshaw also gave his view of the continuing FPC labeling controversy: "Only the extraordinary political climate surrounding FPC would cause anyone to take issue with the labeling procedure which has been proper in situations in which there is far less evidence of freedom from toxicity to man than in the present case."[30]

The major issue remaining with the FDA, however, was the one-pound package requirement, which effectively prevented any serious entry of the food industry into the FPC market since the product could not be used in manufactured foods. At their meeting with the FDA on May 24, 1968, Alpine Marine Protein Industries expressed particular concern with this restriction, and the company indicated its plans to submit an application for a temporary permit to test market a standardized food containing FPC for consumer reaction. It was concluded that macaroni would be preferred for fortification, because it was nonperishable and would be well suited for overseas shipment, and that the FDA would not object on this basis.

In late July, Alpine followed through by formally petitioning the FDA for removal of the one-pound package restriction and limitations on the use of the product by children under eight years old. No new evidence was presented, but the company advanced a number of arguments:

The restrictions were arbitrary, unreasonable, and unduly precluded the marketing of FPC.

Safety was not a basis for the restrictions. There was more inherent danger in the unsupervised use of the additive by the *consumer* than by a food manufacturer.

There was a need for commercial use.

FPC was wholesome.

Any possibility of consumer deception could be prevented by appropriate label statements. Present general food labeling requirements were adequate to insure consumer protection.

The day following the submission of the Alpine petition, July 30, 1968, a meeting was convened in the office of Congressman Hastings Keith (Massachusetts) to discuss the current limitations on the development of a domestic market for FPC. Among those present at the meeting were Wilbur Cohen, Secretary of HEW; Edward Wenk, Director of the National Council on Marine Resources and Engineering Development (MSC); and representatives of AID, BCF, and HEW. Secretary Cohen was informed of the need to remove the one-pound package restriction before development could proceed, while Dr. Wenk made note of the fact that the original regulation had been issued *provisionally* and should now be reviewed—some eighteen months later.[31]

On August 16, the FDA replied to the Alpine petition, finding it inadequate due to the lack of data supporting the proposed modification. The regulatory agency demanded information on all planned uses for FPC in food, as well as proof that the use of whole fish would not be deceptive to the consumer. The Alpine-FDA dialogue continued in a letter of September 23 from Alpine's attorneys, Markel, Hill, and Byerly. They cited the unavailability of the requested use information since, according to the FDA ruling, FPC could not be used in manufactured foods—a "catch 22." In addition, data was presented from Dr. George Graham of the Johns Hopkins University, indicating that it would be virtually impossible for children under eight years old to ingest more than 20 grams per day of FPC—even if 100 percent of their food was composed of FPC at the recommended dietary limit.

The Alpine effort to obtain the removal of the one-pound limitation ended after a response from the FDA on November 22, stating that they still did not consider the petition to be in a form acceptable for filing. They requested additional information on conditions of proposed use, levels of maximum consumption, types of beneficial effects, actual FPC samples with proposed labels, and further evidence that the consumer would not be deceived. The FDA position seemed absolutely firm, and consequently Alpine discontinued its efforts.[32]

It had become evident that even the FDA was interested in putting an end to these continually troublesome matters despite their rejection of the Alpine petition. What was needed was a soundly conceived and well-executed petition that the regulatory agency would find it possible to accept. Toward this end, a meeting was held on April 16, 1968, in the offices of Markel, Hill, and Byerly, the Alpine attorneys, attended by the attorneys and representatives of the Bureau of Commercial Fisheries, to discuss possible strategies for the petition. It was decided that, in view of the FDA's resolute position concerning the packaging question, it would be more practical to seek an expansion in the number of approved fish species and *then* seek the removal of the packaging restriction on a "product-by-product" basis. Discussions with the FDA consequently focused on the right types of fish to be included in a petition.

The decision to divert attention from the packaging question to the expansion of the approved species list is susceptible to a certain amount of retrospective criticism. It is clear that the Bureau was particularly aware of the need to expand the approved species list so as to ensure the effective operation of the Experiment and Demonstration Plant on the West Coast. But the decision by the Bureau not to follow through on a policy decision already initiated, namely, on the packaging issue, gave the FDA, weary from eight years of constant pressure on the FPC question, the opportunity to reconsolidate its own policy position. Thus, the BCF ultimately worked in opposition to its own objective and the ruling on packaging was permitted to remain in effect for another four years.

The Alpine executives objected to the proposed concentration on menhaden, herring, and anchovy, which they said would tend to benefit the Canadian companies more than the United States due to the location of the fisheries. But the BCF pointed out that they were responsible for the development of *all* underutilized species, of which the fatty fish represented a large proportion. As a result of this meeting, and further work done at the College Park Laboratory, a series of possible strategies evolved:

1. ask for general approval of a whole class of fish; alternatively, the approval sought could be made increasingly more specific, that is, various orders within classes, or various families within orders, or even various species (menhaden, herring, and so on) within families;
2. ask for approval for all nontoxic, bony fish;
3. ask for approval of all fish that historically have been found to be edible;
4. ask for specific approval for herring and herringlike species and for ground fish;
5. ask for specific approval for herring and herringlike species and for white fish; and
6. ask for specific approval for herring and herringlike species and for bottom fish.[33]

The consensus of opinion was that to seek a more general approval would entail unnecessary delay, due to the comprehensive nature of the request, while to go to the other extreme and request only a specific species was considered too limiting and cost ineffective. The choice lay, therefore, somewhere in the middle, perhaps requesting the approval of a fish family, while steering clear of the sports fish families which could provoke the opposition of the powerful sports fishing interests.

During the fall of 1969, meetings were held between the BCF and FDA to negotiate the exact content of the petition. The FDA raised the possibility of the need to conduct reproductive and feeding studies since it was concerned about the toxicological effects of different species of fish with varying food ing habits. Despite the possibility of further delays, the BCF filed a formal petition on December 24, 1969, requesting the expansion of permissible species to include herring, menhaden, ocean pout, anchovy, and alewife.[34] The rationale for the petition was that the present regulations permitted only hake and hakelike species while other abundant and underutilized species had to be thrown back, which was both wasteful and expensive; and that the successful manufacture of FPC prepared with the IPA solvent extraction process had been achieved using fatty fish.

On April 1, 1970, the FDA informed the BCF director that, due to questions concerning the feeding habits of the proposed species, insufficiency of data to support the representativeness of the species, lack of toxicological data on the different species, and differences between warm/cold and fresh/salt water fish species, the agency was "unable to recommend favorable action on this proposal to provide for a broad spectrum of fish in the production of whole fish protein concentrate."[35] The FDA did offer, however, to proceed with the approval of menhaden and herring which it agreed were safe.

The BCF accepted the FDA offer in early May and by June had provided the requested certification that the species of menhaden and herring used would not differ from that on which toxicological tests had been performed. Concurrently, the NAS Committee on Marine Protein Resources transmitted a letter to the FDA concerning the alleged toxicity of fish viscera. The scientists found the FDA's concern unwarranted on the basis of available evidence,[36] and they recommended selectively approving genera within larger taxonomic grouping, specifically cods, hakes, pollocks, herring, menhaden, alewife, shad, sardines, and anchovies.[37]

Finally, on August 4, 1970, the FDA announced the amendment of the Code of Federal Regulations (Section 121.1202) to provide for the use of herring and menhaden using the IPA solvent extraction method only, and not the VioBin dual-solvent extraction process.[38] Although it was not immediately clear why this distinction was made in the ruling, it may have resulted, in part, from unanswered FDA questions concerning both the capability of the dual-solvent process to deal with the higher lipid content in fatty fish, as well as the recent, inadequate product performance of the VioBin-licensed process at the Alpine Marine Protein Industries plant.[39] In any event, the issue became academic when the VioBin Corporation withdrew its petition for additional species without prejudice on September 23.

THE END OF THE STRUGGLE

Despite the fact that the use of whole FPC had been sanctioned and that the list of approved species enlarged, there was virtually no possibility of attracting private investment in FPC due to the continued enforcement of the bulk packaging limitation. Few companies were likely to be interested in marketing a product which could be sold only at the consumer level (that is, not admixed in a manufactured product), and even then, only in one-half or one-pound packages. The FDA seemed content to ignore the urgings of the White House Conference on Nutrition (and others) for a relaxation of the ruling, and thus the situation remained static into the spring of 1971.

On June 2, however, the FDA (which had unexplainably resurrected the 1968 Alpine Geophysical Associates bulk-packaging petition and published it on February 18, 1971)[40] formally notified Alpine Geophysical Associates that its petition was rejected. It stated three reasons for the denial: (1) no acceptable methodology for regulatory procedures was proposed; (2) insufficient data on levels of mercury, cadmium, strontium, selenium, and arsenic were presented; and (3) the petition did not demonstrate that the proposed use would not result in consumer deception.[41] Since the Alpine FPC plant had long since ceased operations, the company saw little advantage in continuing its efforts. The petition was formally withdrawn on September 30, 1971.

The Bureau of Commercial Fisheries, now renamed the National Marine Fisheries Service (NMFS), continued to maintain a strong interest in obtaining the removal of the obstructionist packaging limitations. However, since the FDA ruling applied only to the *commercial* production of FPC, the role of NMFS as a petitioner for the removal or modification of the regulation was considered inappropriate. Thus, with no petition actively before the FDA, prospects for change looked bleak indeed. One last hope remained, however, namely, to convince the industry trade association, the National Fish Meal and Oil Association (NFMOA), to become involved again.

The NFMOA, and the fishing industry in general, had shown an increasing reluctance to actively associate itself with the federal FPC development program as its problems had compounded. But, at the end of December, the trade group was convinced to make one more attempt to develop a domestic, commercial market for FPC by submitting another petition to the FDA for removal of the packaging limitation. Like the Alpine initiative, the petition requested the amendment of the food additive regulations to allow the manufacture, shipment, and use of FPC under commercial conditions. Unlike Alpine, however, this petition presented evidence concerning the occurrence of protein deficiency in the *U.S.* population and it demanded that FPC be allowed to compete equally on its own merits with other new food products such as textured vegetable protein or single cell protein.

The NFMOA proposed that the maximum use of FPC in manufactured foods should be governed by the total level of fluoride, which, it suggested, should not exceed 8 ppm dry weight. During the winter of 1972, the FDA responded to the latest "assault" on the FPC regulations by demanding more information on heavy metal content (such as methyl mercury, cadmium, strontium, selenium, arsenic, and lead), as well as additional data on proposed uses and labeling.[42]

For most of 1972 the NFMOA worked closely with the scientists and food technologists of the NMFS, College Park Laboratory, to produce an acceptable response to the FDA demands. On November 6, the revised petition with the requested data was ready to be submitted. The petition proceeded through normal FDA channels, and on the recommendation of the Compliance Regulations Policy staff, it was presented for consideration to the FDA commissioner on January 31, 1973. Just when it appeared that the absolute, final hurdle had been cleared, yet another delay developed: the FDA commissioner refused to rule on the petition due to the lack of unanimity of opinion among his own staff. The commissioner ordered that the petition should be held until the agency staff could agree on a course of action.

Ultimately, however, the FDA approval process, which had begun some 12 years earlier with the first proposal for an FPC food standard of identity (in 1961), came to an end on July 17, 1973, with the publication of an order in the *Federal Register* officially removing the packaging restrictions from the commercial use of FPC.[43] The order stated that "these restrictions are no longer deemed necessary provided that the labeling of the manufactured foods containing the additive bears the declaration 'whole fish protein concentrate' in the proper sequence in the statement of the ingredients."[44] The rationale given for the decision was that the name "whole fish protein concentrate" had become so well known that it would no longer be deceptive to consumers, and thus, the need for packaging restrictions was removed. Ironically, the FDA had, in fact, wound up using its own original anti-FPC argument (that the consumer was not aware of the content of FPC and would be deceived by its use) to justify the complete reversal of policy.

The ruling also contained a number of other provisions. The FDA reasoned that, because the fish used were deep ocean fish, there was not significant danger of exceeding the 0.5 ppm tolerance limit for methyl mercury and other heavy metals. But, due to the agency's continuing concern about the effects of fluoride on children's teeth, an 8 ppm (dry weight) fluoride tolerance was imposed on use of the product in manufactured foods. Finally, the ruling also responded to an earlier petition by the NMFS, of June 2, 1971, by adding anchovy to the list of approved species.[45]

CONCLUSIONS

Although the timing of the final FDA capitulation on its regulations regarding FPC may be suspect, there is no evidence available to substantiate the claim that the final removal of the packaging restrictions was withheld until all private commercial interest had terminated. The evidence reviewed here does suggest, however, that the FDA exceeded the scope and extent of its congressional mandate by permitting personal bias and political pressure to figure prominently in the consideration of a product that had received overwhelming scientific endorsement. By the same token, it must be stated that the BCF/NMFS did not, through a number of its actions (or lack thereof), act in a manner that was likely to aid its own cause at many crucial points between 1961 and 1973.

Part of this less than adequate BCF/NMFS response may be ascribed to a certain degree of naiveté with regard to the nuances of gaining approval and acceptance of a novel food product. Unlike the large food processing companies, this type of activity was (and is) obviously not the agency's principal modus operandi. But the other primary cause of the BCF's lack of success with the regulatory agency was the fact that the Bureau was consistently forced to fight alone. The BCF/NMFS never enjoyed the kind of support from the fishing industry that farmers give to the Department of Agriculture, nor did it have the benefit of the kind of well-financed and well-executed attack characteristic of the large corporations. In fact, during the later stages of the program, the BCF/NMFS could not even count on the support of its own department, the U.S. Department of the Interior. The FPC program was, in short, a white elephant that had neither the support of interest groups or bureaucratic advocates, nor the ability to deal effectively with the FDA.

As Arnold Singal has suggested,[46] the Congress did not intend that the Food, Drug and Cosmetic Act be used to limit the advance of technological progress in the food industry, nor did it intend the FDA to act to limit the consumer's freedom of choice.[47] Yet, over and above the poor policy decisions of the BCF/NMFS, the actions of the FDA regarding the decisions on FPC clearly violated both of these tenets. Singal has noted that only the Congress may be able to provide a continuous review of food and drug policy to prevent such future abuse of mandate, since the President does not have the authority and the courts have shown a great reluctance to become involved in such matters.[48]

The interaction between the BCF/NMFS and the FDA raises a number of profound questions not only about the ability of regulatory agencies to rationally regulate the process of technological innovation, but also about the capacity and advisability of public agencies assuming the development responsibility for projects that, due to their cost or inordinate risk, have been shunned by the private sector. When decisions weighing economic considerations against humanitarian or social ones are to be made, attention to these kinds of concerns is critical.

Exhibit 1
Objections to the FDA Order of February 2, 1967

Ninety-nine communications were received within the statutory period provided for objections to the order. Ninety-one of the letters were from individuals, almost all of whom were dairy farmers from Wisconsin objecting as individuals and as farmers required to maintain high sanitary standards in their own product. Four letters specifically alleged adversity, and three actually requested a hearing. A representative group of these communications are included here (see exhibits 2 and 3) as examples of the concern expressed.

Letter No.	Correspondent	Major Concerns
1-62	Individual Dairy Farmers, Wisconsin	1. A filthy product that individual would not eat. 2. Dairy farmers must scrupulously observe sanitation to avoid including filth permitted in fish concentrate by the regulation. 3. As a cheap source of protein, this will replace more expensive milk protein and bankrupt farmers.
63	Great Plains Wheat, Inc., Kansas City, Kansas	1. Oppose use of whole fish flour in American food. 2. Questions value and advisability of using additive.
64-91 & 93	Individual Dairy Farmers, Wisconsin	Same as 1-62
92	Pure Milk Products Cooperative, Du Lac, Wisconsin	1. Members "shocked that responsible government officials should submit to the pressures of a few profit hungry industrialists."

Exhibit 1 (continued)

Letter No.	Correspondent	Major Concerns
92 (continued)		2. In protein abundant America no need to take risks to human health such as recognized in order.
		3. Approval of fish meal will be serious handicap to consumption of products containing it.
		4. Approval will be deterrent to consumption of many healthy foods to which it might be added.
94	South Carolina Dairy Association, Inc., Chester, South Carolina	1. Order against public health and welfare and not in public interest.
		2. Heretofore FDA has consistently held that such a product including heads, fins, tails, viscera, and internal contents of fish and its guts is filthy and unfit for human consumption and it is especially injurious to children.
		3. Recognized as dangerous and injurious to health as indicated by specifications in order.
		4. Dairy industry adversely affected by substitution of fish flour for nonfat dry milk.
		5. Limitations on household consumption by children unenforceable.
		6. Dairy industry put to great expense to avoid incorporation of filth, making fish flour cheaper.
		7. Enforcement of regulation on dairy products more effective than on fish flour.
		8. Should be cleared with USDA.
95	Wisconsin Council of Agriculture Cooperatives Madison, Wisconsin	1. Objects to approval on basis of findings of FDA in F.R. 1/25/62.
		2. Whole fish flour will undermine dairy economy.
		3. FDA has not evaluated use of fish flour protein in human diet over extended period. Health of children placed in jeopardy.
		4. Opens new area of microbiology. Toxins may have detrimental effects on functions of human body.
		5. Public cannot be assured that sick or diseased fish will not be used; contradict requirements for other protein sources.
96	Department of Health State of Hawaii Honolulu, Hawaii	1. Objects as citizen and Food Commissioner.
		2. Would be required to quarantine and seize all such food additives.
		3. Solvent or solvents remove only fat and moisture not filthy or unwholesome food ingredients inclusion of which, under Hawaii law, makes food adulterated and subject to seizure.
97	Dr. Franklin C. Bing Chicago, Illinois	States Issues:
		1. Is designation as Whole Fish Protein Concentrate proper?

Letter No.	Correspondent	Major Concerns
		2. Is the product properly considered as not in conflict with sect. 402 (a)(3)?
		3. Is "hake-like fish" a proper designation?
		4. Is the limitation on use by children reasonable and in interest of consumers?
		5. Does Commissioner have authority to regulate size of packages?
98	Charles M. Fistere Washington, D.C. (on behalf of American Dry Milk Institute)	States issues:
		1. Whether product consists in whole or in part of any filthy substance and is adulterated under sect. 402 (a)(3).
		2. Whether it is reasonable to permit use of heads, tails, and so on.
		3. Whether method described produces adulterated product under sect. 402 (a)(4).
		4. Whether product is adulterated under State statutes.
		5. Whether regulation is unlawful under sect 409 (c)(3)(b) in that use promotes deception and otherwise results in adulteration.
		6. Whether the name FPC is inappropriate.
99	Wisconsin Creameries Association, Madison, Wisconsin	Incorporates same points as Fistere letter.

Source: Adapted from Food and Drug Administration files.

Exhibit 2

Mississippi Valley Milk Producers Association
1919 Second Avenue, Moline, Illinois 61265

June 3, 1966

Dr. James L. Goddard, Food and Drug Administration
U.S. Department of Health Education and Welfare
Washington, D.C. 20204

Dear Dr. Goddard:

We are extremely disturbed by the information that we have concerning the possible legalization of whole fish flour for consumption in this country.

As you know, our dairy industry operates under very stringent regulations. There have been tremendous strides forward in the dairy business in recent years. We believe that we have wholesome dairy products available for consumers today and in adequate supplies.

We cannot understand how anyone could consider allowing the complete grinding of whole fish, including the heads, fins, scales, tails and guts and then allow this to be consumed by human beings. Our packing houses are very careful to keep intestinal content and viscera out of the products that we eat and we can see no difference between fish intestine and those of other animals.

Our dairy farmer members in Wisconsin, Illinois and Iowa are very vocal in opposing the matter. They realize the implications of it on the public health aspects as well as on economics of our business. We urge you to do everything within your power to avoid the approval of this type matter for human consumption.

Very truly yours,

Mississippi Valley Milk Producers Association
Fred A. Hamann, President

FAH:ind
cc: Board of Directors
Iowa, Illinois and Wisconsin Senators

Source: Food and Drug Administration files.

Exhibit 3

October 17, 1968

Dr. James L. Goddard, Commissioner
Food and Drug Administration
U.S. Dept. of Health, Education, & Welfare
Washington, D.C.
20204

Dear Sir:

Your stand against the Whole Fish Flour Issue is a laudable manifesto of the integrity and true Americanism which has characterized your activities wherever the welfare of your countrymen has been at stake.

With the clear vision of men devoted to the obligations of their assignment you fearlessly refused the approval of this scheme because of its unfitness for human consumption. The reckless $8.8 million research fund allocated for developing this project is a pathetic waste of United States money. All thinking men and women can, (without scientific paraphernalia and huge-salaried specialists) discern the filth consequent upon the processing of *"whole fish"* . . . fins, heads, tails, scales, intestines and their contents. The cost of the suitable processing of fish, conformable to the accepted standard of cleanliness would far out-weigh the expenditure involved in the *"dried milk and dried eggs"* products for which the "Whole Fish Flour" is proposed as a replacement—healthwise and economy-wise, neither of which it touches by a finger-tip.

Our farmers, not only Americans, but those also who come from many lands to absorb our customs and culture, continue to till the soil, generously watered and warmed by God's rain and sun producing grains that are the admiration and envy of agriculturists the world over, feeding people and cattle. These cattle supply the milk, which, together with the grain, brings answer to our pleading prayer, "Give us this day our daily bread."

Our hopes, confident and assured, are being placed in your competent, kindly hands, knowing that you will continue to stand unafraid against those sharp operators who are so prone to suggest inferior replacements in foods that are as familiar to us as marbles and jacks.

Today, in this new crisis, it is your privilege and honor of keeping America—American.

Sincerely yours,
Medical Librarian

Source: Food and Drug Administration files.

Exhibit 4
Resolution on FPC of the Association of Food and Drug Officials of the United States.

Report of Resolutions Committee June 18-22, 1967

RESOLUTION

WHEREAS, public announcement appeared in the Federal Register for February 2, 1967 (FR Vo.32, No.22) advising that the product *"Whole Fish Protein Concentrate"* would be included in Federal "Food Additives Permitted in Food for Human Consumption," and

WHEREAS, this product is the same processed "whole fish," including fish heads, tails, fins, scales, and viscera with its content of ingesta and fecal matter—the same product previously identified as the "Whole Fish Flour"—which this Association opposed by Resolutions adopted in Annual Conferences, the 66th, June 17-22, 1962, Hollywood, Florida, and the 67th, June 16-20, 1963, Lansing, Michigan.

Now therefore be it Resolved that the Association of Food and Drug Officials of the United States, in its 71st Annual Conference, June 18-22, St. Paul, Minnesota, in its continued opposition to the degradation of human foods and the food standards of this Country, hereby reiterates the aforementioned Resolutions, and

Be it further Resolved that this Association endorses and supports the Mid-Continental Association of Food and Drug Officials in the Resolution of its Annual Conference, March 13-14, 1967, Wagoner, Oklahoma, reiterating its opposition to the use of filthy materials, or materials otherwise unfit for food, as ingredients in foods regardless of whether they are introduced under the name of foods, food ingredients, or food additives.

RESOLUTION

In the interest of uniformity in the area of packaging and labeling it is hereby resolved:

Until the Federal regulations under the Fair Packaging and Labeling Act are finalized no action be taken to amend present laws or regulations or pass new laws or regulations in the area of packaging and labeling, except such changes in laws as are necessary to authorize the dual declaration of the net quantity statement and the defining of consumer and non-consumer packages.

Be it further resolved that copies of this resolution be forwarded to all State and municipal regulatory agencies for their consideration.

Source: Food and Drug Administration files.

Chapter 5
The Evolution of Standards for Acceptable Levels of "Filth" in Food

FILTH AND THE CONSUMER

Concern about the presence of "filth" in whole fish protein concentrate has been manifest throughout the involvement of the U.S. Food and Drug Administration. From the initial attempt to have the product approved as a food in the United States through the major efforts to gain its acceptance as a food additive, to the final struggle to have the bulk packaging limitation removed, various versions of the same argument against FPC were reiterated.

Initially (from 1961 to 1967), the FDA position was that the adulterated contents of FPC caused it to violate Section 402(a)(3) of the Food, Drug and Cosmetic Act of 1938, which stated that a food shall be deemed adulterated "if it consists in whole or part of any filthy, putrid or decomposed substance, or if it is otherwise unfit for food."[1] After the product was approved as a food additive in 1967, however, the filth argument reappeared, in a modified form, through the unprecedented imposition of the one-pound package limitation. The preamble to this ruling stated, in part, that "individual consumers are entitled to the opportunity of choice in the matter of whether they desire to include this item in their diets"—a rather thinly veiled reference to the possible undesirability of the product—and that packaging in units larger than one pound could not be authorized unless it was demonstrated "that the proposed use of the article will not promote deception and will promote honesty and fair dealing in the interest of the consumer."[2]

On first analysis, it may seem extraordinary that an agency of the federal government, known for its rigorous empirical standards, would spend twelve years basing its principal objection to a product on subjective considerations of aesthetic acceptability. But a closer examination of the Food and Drug Administration—in terms of its origin, enabling legislation, and historical mission—reveals the precedent for such a position. The FDA perceives its fundamental mission to be that of benevolent protector of the innocent, unknowing consumer. "The consumer is an amateur; he knows next to nothing about the multiple of processes and multiplex of ingredients which lie implicit within a commodity. He cannot judge whether the consumption of an article contributes to his well being."[3] This paternalistic view of the consumer has also been upheld repeatedly by various judicial interpretations of the Food, Drug and Cosmetic Act, for example, in the *Marmola* case:

> The Federal Food, Drug, and Cosmetic Act was not made for experts, nor is it intended to prevent self-medication. The purpose of the law is to protect the public, the vast multitude which includes *the ignorant, and unthinking, and the credulous* who, when making a purchase, do not stop to analyze.[4]

and again in the controversial *Dotterweich* ruling:

> Congress has preferred to place it upon those who have at least the opportunity of informing themselves of the existence of conditions imposed for the protection of consumers before sharing in illicit commerce, rather than to throw the hazard on *the innocent public who are wholly helpless.*[5]

Since more and more of the food consumed each day by the average individual is of a complex, manufactured variety, prepared in some anonymous location, the consumer is no longer able to observe, much less determine, either the nature or the purity of the products eaten. New methods of distribution have also made the relationship between producer and consumer entirely impersonal. The consumer, therefore, turns to government to provide protection through the enactment of statutes, such as the food and drug laws, that prohibit injurious practices and that are enforceable by government action. The government is thus made responsible for inspecting manufacturing plants, a function that the consumer had found impossible to perform. Labeling requirements are set by the government to furnish the consumer with the information needed to make a rational selection among packaged products. Most important of all, the government enforces its regulations by suits at public expense.

ORIGINS OF THE U.S. FOOD REGULATIONS

Many of these aspects of the FDA mission had their origins long before the creation of such an agency was even considered. In early agricultural societies every family produced its own food supply or obtained it in local markets. Foods were few and simple. With foods prepared at home, carelessness in sanitary matters would only affect the preparers themselves.[6] Raw and crude food stuffs were the chief commodities available in public markets. Grains and spices were not ground. Vegetables and fruits were sold fresh or dried. Prepared foods, as we understand that term today, were very rare. Only such foods as honey, ales, wines, and oils offered the possibility for adulteration.[7] As commerce developed, opportunities to profit through adulteration also increased.

Food regulations were included in the Laws of Moses written over two thousand years ago. Requirements for just weights and measures can be found in Deuteronomy. The Book of Leviticus includes elaborate provisions to ensure the purity of the food that the Hebrews consumed.[8] Early Greece and Rome provided for inspectors to guard against adulteration of their wines.[9] Governments began to regulate bread baking, milling, slaughtering, and wine and beer making as these became recognized trades in the Middle Ages. The Magna Carta, for example, provided for a standard measure of wine, ale, and corn in thirteenth-century England.[10] A 1597 English statute provided "that if any shall bring into any haven, port, creek or town of this realm any salt fish, or salt herrings which shall not be good, sweet, reasonable, and meet for man's meat, and shall offer the same to be sold, then all and every person, owners thereof, shall lose and forfeit to our Sovereign Lady all said unreasonable fish."[11]

The Industrial Revolution brought advances in food processing that allowed inedible things to be made edible. New ways were discovered to preserve perishable foods. Food wastefulness could be curtailed by developing new by-products. Many other things could be accomplished with raw materials lacking in uniformity.[12] As food processing moved from the home to large factories, the consumer became dependent on the general market. No longer could one be sure that food was prepared in a sanitary manner. However, no general food law was passed by the U.S. Congress during the nineteenth century, even though the condition of the nation's food supply was far from satisfactory. The following statistics were gathered in 1880:

1. 73 percent of the milk of Buffalo, New York, was watered;
2. 69 of 171 samples of ground coffee collected in New York were adulterated;
3. 71 percent of olive oils examined in New York and Massachusetts had been mixed with cottonseed oil;
4. 46 percent of candy samples from Boston contained mineral pigments, chiefly lead chromate;
5. New Hampshire's State Board of Health found only 21 of 41 samples of maple sugar were genuine.[13]

Federal action became necessary because state laws passed in the middle of the century were inadequate to control these practices.

The first general food law for the United States was proposed by Senator

Paddock of Nebraska during the 51st Congress of 1891. No favorable action was taken on the measure. Similar attempts were made in the House between 1901 and 1905, but proponents of these bills could not get them by the special food industry interests entrenched in the Senate.[14]

A long campaign of public agitation finally resulted in the enactment of the 1906 statute. Dr. Harvey Wiley, the chief chemist for the Department of Agriculture, formed a "poison squad" from members of his staff. This group ate questionable foods being sold to the public and reported the state of their health daily to the press. Upton Sinclair dramatized the unsanitary conditions prevailing in the Chicago meat packing plants in his widely read novel, *The Jungle.* An effective educational campaign was conducted by the General Federation of Women's Clubs.[15] As a result of these pressures, President Theodore Roosevelt included a request for legislation governing the misbranding and adulterating of foods in his message to Congress on December 5, 1905. After each house of Congress had passed its own version of this legislation, a conference committee was appointed. Both houses agreed to its report, and the first general food law was passed by Congress on June 29, 1906.[16]

Just how far Congress meant to go in regulating food products to protect consumers was still not clear. The immediate aim of Congress was the halting of outrageous practices that threatened the health and welfare of consumers and the competitive position of reputable manufacturers. Concerns about the preservation of consumer choice and continued growth of the food industry remained in the background.

The 1906 law as enacted soon revealed many gaps in its protection of consumers. The growing trade in cosmetics was outside the scope of the act. Foods sold under proprietary names and those labeled as compounds, imitations, or blends were exempted. The Food and Drug Administration found it difficult to prove that goods were misbranded without the authority to establish standards of identity.[17] By 1930, proponents of change in the food and drug laws decided to draft a new statute rather than make patchwork amendments. A committee was organized to perform the task and, by June 1933, a prepared draft was ready to be submitted for legislative action as Senate Bill 1944.

Not many changes had been made in the provisions concerning adulteration. Section three of the proposed act defined a food as adulterated "if it consists in whole or in part of any filthy, putrid, or decomposed substance."[18] The 73rd Congress failed to act on the bill, however, due to fears about the discretionary powers to be granted to the FDA. The measure was reintroduced in the next session, at which time a comma was inserted in the adulteration section after the word "substance" and the clause, "or if it is otherwise unfit for food," was added. No reason for this addition appears in the record. Either the secretary of agriculture had asked for a broader statement to help with tough cases, or the draftsmen had thought it wise to add a residual clause. Although different versions of the bill passed each of the Houses, a dispute concerning the agency's jurisdiction over false advertising prevented the bill's enactment.

A variety of developments paved the way for the enactment of an almost identical Senate bill during the next session. The public had become aroused following the sale of Elixer Sulfamilimide, a drug compound containing a deadly poison, on the open market during the autumn of 1937. Secondly, a controversy over control of advertising between the FDA and the FTC was settled by the Wheeler-Lea Act.[19] Finally, a major change was made in the judicial review provisions of the Senate measure to satisfy some of the bill's most vocal critics.[20] With many of the bill's controversial issues settled and confronted by pressure for stronger drug legislation, the members of the 75th Congress were prepared to agree on the terms of a new food and drug statute that became the Federal Food, Drug and Cosmetic Act of 1938.

INTENT OF SECTION 402 (a) (3) OF "FILTH" REGULATIONS

The preceding analysis is illustrative not only for the light it sheds on the history of attempts to regulate adulterated foods in the United States, but also for its consideration of the very notion of filth in foods. Legislative history provides the Food and Drug Administration with few clear indications of congressional intent. The need to protect the consumer's physical safety was emphasized a number of times in the legislative debates. These debates also reveal the concern of Congress for the prevention of deceptive practices made possible by advances in food technology. A by-product of this goal is the protection of the ethical producer from unscrupulous competitors who want to profit from cheapened products.

No doubt Congress also meant in some way to protect the sensibilities of the American public. But protection of consumer sensibilities involves some difficult aesthetic considerations: "Every culture prescribes for its members the meaning of food substances, methods of preparation, time and style of consumption and the like."[21] People in the United States, as elsewhere, often exhibit inconsistencies in their attitudes concerning food. Foods that are enjoyed by large segments of the public use components that are aesthetically unacceptable to other segments.[22] No satisfactory chemical index of filth has ever been established for our culture. As a result, the FDA faces a charge of paternalism whenever it acts to protect consumer sensibilities.

FILTH IN FOOD

"The human race instinctively associates the satisfaction of appetite with pleasure and the esthetic quality of desirable food. Strange, unusual, and unfamiliar foods often are distasteful to the point of complete rejection. This reaction undoubtedly grew out of the experience that eating promiscuously could be hazardous business and that it was safer to stick to the tried and true. A lesson undoubtedly learned early was that many foods quickly lost their wholesomeness. What could be eaten with impunity one day could be disgusting, if not actually toxic, the next. Decomposition and putrefaction were respected warning signs long before the days of Louis Pasteur. It is understandable that the nature of spoilage and its association with unclean environments led to the general recognition of filth as a factor contributing to the unwholesomeness of contaminated foods."[23]

But what is filth? The word itself is derived from "foul" which, in turn, originated from the Latin verb *putere:* to rot, to be decayed, to stink. It seems therefore, that at least originally, an olfactory sensation was an important part of the concept of filth. It is likely that initial concepts of filth were associated with material suspected to contain substances resulting in disease. Perhaps this may explain why "filthy" material is most generally associated with decayed, spoiled tissue, and excreta, that is, metabolized food. In a study of the components of human feces, Goldblith and Wick (1960) reviewed the literature and found that the knowledge of human fecal components was very incomplete, that detailed knowledge of the chemical nature exists in only a small proportion of fecal components, and that these components have rarely been isolated in any quantity.[24]

The following are examples of esthetically objectionable foreign matter, that is, filth, that is present in common foods and that for practical reasons is tolerated to a degree:

fecal extract remaining in cow's milk after filtration for the removal of manure fragments;

excreta of fish and birds in streams and reservoirs furnishing potable water supplies;

insect fragments and excreta in flour, corn meal, cereals, grains, cocoa, coffee, nuts, fruits, and vegetables;

rodent excreta and urine in wheat, corn, and other grains;

intestinal contents in sausage casings, shell fish, crustaceans, roe (cavier), and canned fish, such as sardines, anchovies, and herring;

mold in tomato products, berries, raw fruits, and vegetables; and

worms (including their intestinal contents) such as trichina in pork.

It is recognized that such forms of extraneous matter are unavoidable despite sanitary manufacturing practices, including processing steps such as filtration and centrifugation designed to reduce contamination to a minimum. Because no health hazard accompanies the presence of these residues, enforcement agencies have established levels of tolerances that in some cases are published and in others are tacitly understood to exist.[25]

"The question arises whether visceral organs per se, that is, aside from the contents of the gastrointestinal tract, are esthetically objectionable as sources of food. To dispel this generalization, one need only cite a few of the organs used as food by civilized man; namely, tripe, liver, kidneys, heart, lungs, sweetbreads, and intestines (as sausage casings), not to mention blood and brain. The fact that we do not generally eat certain viscera, like mammary glands and gonads, which are consumed by peoples in other regions, may be explained by the greater availability of animal foods in our economy and hence our opportunity to indulge in dietary preferences to which we have become adapted. Obviously these visceral organs are neither unwholesome nor distasteful to those who have not acquired inhibitions toward them."[26]

The choice of foods is predicated on considerations of nutrition, safety, aesthetics, and economics; climatic and geographic factors also play an important role. Foods consumed by certain ethnic groups may be regarded with disgust by others. Americans, for instance, reject fermented tomatoes but accept soy, a sauce from fermented soybeans. The Japanese relish a condiment prepared from spoiled whole fish at which Americans would probably shudder. Yet, many people enjoy eating yogurt, or Roquefort, Gorgonzola, and Stilton cheeses that are, in effect, cultures of molds, while rejecting other varieties of cheese that become moldy with storage. Oysters are often used as another example of a food that many Americans eat, knowing they are consuming the whole organism (guts and all) while many individuals refuse this mollusk for the very same reason. Ants, grasshoppers, locusts, caterpillars, bees, and other insects (including their intestinal contents) and different varieties of larva are all consumed as food today in some regions of the world. Such foods as octopus, snake meat, and dog meat are repellent to some of us, while milk, beef, and pork are objectionable and even forbidden to others.

Thus culturally oriented food patterns may result from natural or artificial conditions to which ethnological groups adapt themselves. Perhaps intuitive acceptance by different peoples of food forms from particular sources makes nutritional sense and has precluded perceptions of filth being associated with some items, for instance, soya as a complement to rice which is a staple food item for many people. The rather important point here, however, is that it would be difficult to randomly select a national index of filth, that is, some specific compound or compounds that might be considered by all or even an unknown majority or minority, as "filthy." In any case, we seem to perceive "filth" as such only when this "filth" is present in sufficient concentration and commonly recognizable. There is also some indication that attitudes toward that which is commonly recognized as filth may be influenced by media exposure, for example, changes in public attitudes after publication of the fact that the astronauts were consuming parts of their own wastes.[27]

FILTH IN FISH

Is whole fish unwholesome? "Insofar as its nutritional values are concerned, fish ranks among the best sources of food known to man. As with all animal foods the quality of the protein and the proportions of individual vitamins and other nutrients varies in different organs. Moreover nutritional values are affected in varying degrees by processing conditions."[28] Yet it is well known that "raw fish provides a highly nutritious milieu for the propagation of microorganisms. From time immemorial dessication and salting, and in later years, refrigeration and freezing, have been employed to preserve fish against bacterial spoilage. In recent times, antibiotics and other chemical agents have shown promise of usefulness in areas where more conventional means of preservation are impracticable. Bacterial contamination of fish is largely the result of exposure and handling after the catch and only to a lesser extent is it of endogenous origin. Coliform bacteria, such as Escherichia Coli, whose presence is usually considered to indicate fecal contamination, have been identified in the intestinal contents of many varieties of marine fishes. There is good evidence, however, that these microorganisms are derived from polluted waters and are not part of their normal intestinal flora."[29]

"Certain types of bacteria which may contaminate raw fish or fish products have been known to cause food poisoning. Unlike warm-blooded animals, fish do not suffer from bacterial infections of the Salmonella or Shigella type. The presence of these food poisoning bacteria has been infrequently reported. It is believed to arise from pollution, usually of fresh or tidal rather than deep marine waters, and in tropical rather than temperate zones. Whether or not intestinal contents are present in whole fish at the time of processing is dependent on how soon the fish are landed after the catch and whether flushing of the tract has occurred, while holding, before conveying into the processing plant. In this regard practices vary in different fishing areas and with different species of fish. Unsanitary handling on board ship, especially in warm climates, has been claimed to be the major factor in bacterial contamination. Prompt and continuous refrigeration up to the point of heat processing practically eliminates any health hazard."[30]

With regard to the specific question of whether a protein concentrate made from whole fish would be more heavily contaminated than one made from eviscerated fish, two samples of FPC were examined by B. L. Oser of the Food and Drug Research Laboratories (see exhibit 1). One was prepared from whole fish and the other from fully cleaned fish fillets. Both products were fine, free-flowing powders, the former having a tan color, the latter being nearly white. Both were essentially odorless, having at most a faint fishy odor. The samples were submitted to the Food and Drug Research Laboratories through the offices of the National Fisheries Institute, Inc.

"As a part of this study, observations were made on viscera obtained by

Exhibit 1
Bacteriological examinations of Two FPC Samples

Sample		Results
Whole FPC	Total plate count	400 per gm
	Coliforms	Less than 5 per gm
	Salmonella	None found
	Shigella	None found
Filet FPC	Total plate count	550 per gm
	Coliforms	Less than 5 per gm
	Salmonella	None found
	Shigella	None found

It is apparent from this table that no significant differences were found with respect to the bacterial content of these two samples. The bacteria counts were well within the tolerances established by the Food and Drug Administration.[31]

dissection of commercially canned whole sardines, six brands of which were purchased on the open market. Even in these specimens, it was not possible to identify microscopic structures which could be said to characterize intestinal contents. Several specimens of whole, fresh frozen menhaden were also dissected and examined. Insofar as could be determined, the digestive tracts of these fish were essentially empty at the time of examination. From these limited observations, it would seem that criticism of FPC based on the alleged presence of 'intestinal contents' is more theoretical than real. Furthermore, the practicability of a bacteriological or macroscopic index of the presence or extent of 'contamination' with intestinal contents appears to be highly questionable."[32]

JUDICIAL INTERPRETATIONS OF FILTH

Originally, the courts defined adulteration in the same terms as its usual dictionary definition.[33] Proof was required of an actual addition to the original substance by human agency. This is no longer necessary. If the adulteration has been added by nature and is contained in the article to be shipped, it is adulterated under the law.[34] The meaning of the word "filthy" has not been confined to a scientific or medical definition. The courts have ruled that Congress intended that the word be given its usual and ordinary meaning, and thus it becomes a condition to be determined in the light of everyday experience.[35] A number of cases have upheld the Agency's right to rule a product filthy on aesthetic grounds alone.[36]

The courts have sought to limit the government's expansion of the phrase "unfit for food." The government argued in *United States* v. *667 Cases of Canned Herring Roe*[37] that the product was adulterated because it contained parts of the viscera, stomach, intestines, and other disgusting matter from the fish. There was no claim that the product was unhealthy. Instead, the government claimed that it was unfit for food because it contained matter that made it unattractive in appearance. The court refused to accept this argument. It perceived a vast difference between extraneous matter and matter that remained as a result of lack of care in separation. The latter, although unattractive, was not sufficient for a finding that the product was unfit for human use.

In an effort to expand its discretion, the Food and Drug Administration has tried to persuade the courts to accept a liberal interpretation of the words "filthy, putrid, or decomposed." An interpretation of this sort would allow the administration to enforce whatever tolerances it deemed appropriate for a particular product. The courts have not always agreed. In *United States* v. *Gerber Products*,[38] the court concluded it was not the intention of Congress to include as a criminal offense the presence of filthy, putrid, or decomposed matter in such infinitesimal and inconsequential quantities that even the highest degree of care could not eliminate.

ADMINISTRATIVE DISCRETIONARY POWER

The Food and Drug Administration has sought to acquire the broadest possible judicial interpretation of the Food, Drug and Cosmetic Act in order to handle the large number and variety of cases which come before it. Toward this end the agency may assume extreme positions hardly justified by the wording of the Act. Thus, original congressional intent may periodically become secondary to the agency's particular aim of obtaining more effective regulatory power.

An examination of the court decisions under Section 402 (a) (3) reveals that the FDA has been successful in obtaining the power that it desired. As suggested earlier, the courts have not restricted the meaning of the terms of this section to a narrow scientific meaning; even aesthetic considerations may be protected by its terms. The agency has also been given considerable latitude in dealing with "questionable" products as well. Although the FDA has the power to condemn foods containing minute quantities of adulterated matter, it also has the discretion to allow many of these foods to be marketed. Food and Drug officials have admitted that it is commercially impracticable in many instances to produce and market foods that do not contain some small amounts of filth.[39]

For many years the agency has maintained for a large number of products a confidential list of tolerance levels known as "Defect Action Levels" (such as, maggot eggs or rodent excreta in grain, cow feces in milk, and so on). Action is taken only when filth exceeds the level that the FDA considers to be obtainable even with the use of the best possible manufacturing or production techniques. It is significant in this regard that, when the FDA finally made public its Defect Action Levels, there was a tremendous outcry from the press and consumer groups that the standards were not rigorous enough!

Second, Section 306 of the Act gives the administrator the authority to refuse to take action on minor violations "whenever he believes that the public interest will be adequately served by a suitable written notice or warning."[40] Two doctrines developed by the courts can be used by the administrator to allow exceptions. Decomposition can be defined as a state rather than as a description of a process that affects all foods. By means of this reasoning, an administrator can refuse to condemn a food unless decomposition has proceeded so far that the entire product has reached a state of decomposition.

This doctrine cannot be applied as easily to the question of filth which, unlike decomposition, may not be measured in degree. "When filth is present in a product, the court cannot find that there is no filth, and a finding of minute quantities of filth will not exempt the product from the statutory language."[41] The courts, therefore, have used the doctrine of *de minimis* to prevent unduly stringent rulings in cases where the amount of filthy matter is very slight.[42] In these cases, the court simply ignores small amounts of contamination that it believes are impossible to eliminate.

There are certain limits, however, to the amount of discretion that the courts will allow an administrator. It has ruled that any regulation passed in furtherance of the aims of the statute exceeds the legitimate bounds of administrative regulation if it does not operate fairly and reasonably with respect to the producers or distributors of the article as well as to the consuming public. Occasionally the courts have intervened to prevent the Food and Drug Administration from carrying its interpretation of the Act too far. In the *62 cases of Jam* case, the FDA took the position that the product, even though it was clearly labeled, would mislead the public. The court warned that "in our anxiety to effect the congressional purpose of protecting the public, we must take care not to extend the scope of the statute beyond the point where Congress indicated it would stop."[43]

The few cases that do limit the authority of the Food and Drug Administration in specific situations do not limit it very much. The Administration is still allowed broad discretion in applying the statute. The producer of a new food product will thus be dependent on the Administration's approval, recognizing that the chances of prevailing in court are slim. As one prominent food and drug lawyer has suggested, an attorney for such a food producer has two problems: one is to reach an opinion on the basis of the law, and the other is to consider the likely FDA position. "It is difficult to take the FDA to court because the courts equate the FDA with God, motherhood and the American flag."[44] The burden of proof, therefore, falls most heavily on the plaintiff in any such proceeding.

The consequences of unchecked administrative discretion under Section

402 (a) (3) can be quite severe for companies that intend to market a new food product. First, the manufacturer may subject itself to a large amount of unfavorable publicity if it loses a test case.[45] Second, the company may be exposed to the Act's criminal penalties.[46] According to the *Dotterweich* case, even corporate officers and agents may be held personally liable for criminal sanctions. Under the Act's broadly construed provisions, the introduction of an adulterated product in interstate commerce is itself sufficient to expose the manufacturer to criminal prosecution; the FDA is not required to prove intent on the part of the manufacturer.[47]

FUTURE PROSPECTS

The Federal Food, Drug and Cosmetic Act has been the principal guiding statute for the evaluation of purity in food products for over thirty-seven years. During this period, a large and influential body of juridicial precedent has been established for the FDA's interpretations of what constitutes unacceptable adulteration. It is important to draw a distinction, however, between administrative decisions and/or judicial interpretations that have been based on rigorous empirical analysis versus those that have been promulgated largely on the basis of subjective, aesthetic considerations.

Two trends are increasingly apparent with regard to this dichotomy. On the one hand, continued improvements in the state of analytic technology have facilitated greater and greater accuracy in the detection of toxic or "filthy" substances in proposed new foods.[48] This point was emphasized in testimony before the House Committee on Science and Astronautics by Dr. Herbert Ley, of the Food and Drug Administration: "We are in a period of drastic expansion in our knowledge of effects of chemicals on the human body. New analytic technology has made us aware of levels of contamination which would have been undetectable several years ago."[49] While these scientific advances are laudable, it is also worthwhile to consider whether, at some point in the future, the instruments used in analysis might not become sensitive to the trace presence of filth or toxicity at levels that neither would affect the human body nor could be economically reduced by food processors.

The other significant trend relates to the acceptability of declaring a food substance unacceptable solely on aesthetic grounds. In an era when even farmers in the United States are experimenting with feeding animals, such as chickens, with parts of their own wastes (a process known as copraphage), it has become increasingly difficult to prohibit the use of a food or food component based on a priori societal proscriptions. It is particularly instructive, in this regard, to note that the FDA's Office of the General Counsel is also aware of the growing need to reconsider and modify its position on this issue. In a recent public statement a member of that office, Stephen H. McNamara, acknowledged, "Our traditional perceptions of 'filth' and 'unfit for food' may be expected to change as the world's food supply shrinks in relation to the world's population. Aesthetic considerations increase costs and reduce available food supplies."[50]

Mr. McNamara indicated that the FDA is also troubled over the rationale for withholding needed food substances simply because they are without traditional precedent as sources of human nutrition: "As time goes by, it is not at all inconceivable that the agency will be petitioned to permit useful and nutritious foods for human use which will seriously challenge our traditional concepts of filth and fitness for food."[51] Perhaps the ultimate irony in this situation is, as the FPC program clearly demonstrated, that the effects of an FDA action of this sort are felt not only in the United States, where the agency has official authority (and where nutritional need is limited to certain narrow population groups), but also in developing nations where the need is overwhelming but where a combination of nationalistic pride and conservative attitudes toward food (which often accompany its scarcity), may cause the rejection of a nutritionally acceptable product simply because it has failed to receive FDA approval.

The combination of improved analytic capability and continued food prohibitions based on aesthetic grounds serves only to further restrict sources and supplies of food at the very point in time when the opposite situation should be pursued. "As we insist upon 'purer' and 'cleaner' food, rejecting that which does not meet our standards, we thereby reduce the available food supply and increase its cost. Where safety is not a factor, at some point considerations of aesthetics must give way to considerations of human need."[52] It is, indeed, encouraging that the FDA is beginning to admit the need to modify its position on filth and food aesthetics in view of current world food realities. Yet, the body of legal and administrative precedent is large and, as a result, policy changes very slowly. Perhaps, with the aid of congressional modification of the Food, Drug and Cosmetic Act, both the courts and the Food and Drug Administration will be inclined to consider the wider context and social impact of their rulings on novel food products such as FPC.

PART III
INTERNATIONAL FPC DEVELOPMENTS

Chapter 6
FPC in Canada: Government Sponsored Research and the Risks of Privately Funded Development

The concept and history of FPC are inextricably intertwined with Canada: some of the earliest and most critical scientific and engineering research problems connected with the development of a manufacturing process for fish protein concentrate were identified, investigated, and solved there. The interest and confidence with which the Canadian government viewed development of FPC was reflected by the Canadian Minister of Fisheries, the Honorable J. J. Robichaud, in opening the Conference on Fish Protein Concentrate in Ottawa on October 24, 1967: "I think one could safely predict we may well be using the words fish protein concentrate as part of our everyday diet vocabulary before too long. We here in Canada are witnessing the greatest stride in the development of our fisheries."[1]

Since the early 1950s work in Canada to develop the product was supported almost exclusively by the Canadian Government, in particular the Fisheries Research Board of Canada, operating under the aegis and with funding of the Department of Fisheries.

Government support of the Canadian fisheries has been a long-term concern since the fishing industry has traditionally occupied an important place in the country's economy. In certain parts of the country, especially in the Northeast, fishing represents almost the sole source of livelihood for the coastal population. For this reason the Canadian government was willing to initiate and provide considerable technical, financial, and administrative assistance to enterprises that promised to develop and support the fishing industry.

FISHERIES RESEARCH BOARD OF CANADA INVOLVEMENT (1950-1971): PIONEERING EFFORTS

Reflecting the categoric importance of fishing to the economy, the Department of Fisheries of Canada, as a major government department, is headed by a cabinet-level Minister of Fisheries. Under the Department, the Fisheries Research Board (FRB) is responsible for conducting scientific and engineering investigations designed to further and support the Canadian fishing industry. All of the pioneering developments on FPC done in Canada were conducted at the Halifax Laboratory of the Fisheries Research Board where work began in the early 1950s and where it continued until the early 1970s. Interest in producing FPC was stimulated at the Halifax Laboratory by the appearance of Fiat Report No. 520 in 1945. This report had been prepared by the Office of the Military Government of Germany and described the manufacture of a synthetic egg white called "Wiking Eiweiss" which had been produced in Germany from cod and shrimp muscle to supplement the German protein resources during the war.[2]

The first attempt to manufacture FPC in the Halifax Laboratory was made in the early 1950s by a member of the staff, A. Guttmann, but the process was complicated and costly and was eventually shelved.[3] Serious work to develop a defatted and deodorized FPC, suitable for human consumption, began in Halifax in the late 1950s, probably with the intention of marketing the product in Canada where a bland product was deemed essential for acceptability. Two processing routes were chosen to reach this goal: one used enzymatic hydrolysis and the other solvent extraction. Results obtained in experiments with the latter procedure rapidly proved to be more promising, so that most subsequent process developments in Canada (and, as a result, most of the work of the U.S. Bureau of Commercial Fisheries [BCF]) employed this type of operation.

Solvent extraction with isopropanol as the major organic solvent was first successfully demonstrated in Halifax in 1954, with fish roe as starting material.[4] Until 1957, work in Halifax proceeded on a laboratory scale, but during 1957-1958 the process was adapted for pilot plant production and a small pilot operation was assembled. In the course of the next few years hundreds of 25- to 100-pound samples were prepared in this facility and were sent out, free of charge, to government research institutions and private industries in many parts of the world. These samples were valuable since they were known to have been prepared under carefully documented processing conditions and from known raw materials. They were used by scientists of the U.S. BCF and others as basic material in attempts to determine product composition, method of analysis, effects of processing variables on product characteristics, suitability of the product for incorporation into various foods, and so forth.

These Canadian achievements, supported almost totally by the Canadian

Department of Fisheries, stimulated the conduct of detailed investigations undertaken in the late 1950s by the U.S. BCF and greatly assisted Swedish FPC development. In Canada the FPC research finally led to the construction of the ill-fated FPC plant in Canso, Nova Scotia, by Cardinal Proteins, Ltd., discussed later in detail.

FPC research in Canada remained centered at the Halifax Lab and continued to be supported by the Canadian government until the early 1970s. There is no indication in the scientific and technical literature of the 1950s and 1960s that much work on FPC was done by either the Canadian food and feed industries or by Canadian universities. This observation stands in sharp contrast to FPC development in the United States where private industries, academic institutions, and government agencies were active in this field.

With the exception of the advances made by the VioBin Corporation in the United States, working on a different process (see chapter 3 and the appendix to this book), almost all of the important early findings in Canada and the United States (including the use of isopropanol-water mixture for the extraction of lipids from the raw material, the introduction of the phosphoric acid wash prior to the solvent extraction step) originated in Halifax. It was here, too, that efforts were made to utilize filleting line scraps, underutilized species of fish (such as skate and dogfish), fish press cake, and fatty fish species such as herring.[5]

There can be little doubt that the willingness of the staff at the Halifax Laboratory to share their findings and products with the rest of the world had profound effects upon the overall course of the FPC development. It was also from the Halifax Lab that FPC samples prepared from different species were supplied to the Canadian Food and Drug Directorate (the regulatory agency comparable to the U.S. FDA) in the effort to obtain formal approval for the manufacture and sale of FPC in Canada.

DEPARTMENT OF INDUSTRY, TRADE AND COMMERCE INVOLVEMENT (1965-1971)

Like the Department of Fisheries, the Department of Industry, Trade and Commerce is a major government ministry, headed by a cabinet-level minister, which was created "to promote the establishment, growth, efficiency, and improvement of manufacturing industries in Canada . . . to develop unrealized potential, to cope with exceptional problems of adjustment arising from changes in the import/export pattern or in the domestic market. The Department is responsible for promoting the development and use of modern industrial technology in Canada and for improving the effectiveness of research carried on by the Government of Canada."[6]

The Department became involved in FPC development at the urging of Dr. Bernd Weinberg, chief of Agriculture, Fisheries and Food Production, one of the Department's ten branches. In the course of work he had undertaken on problems facing the Canadian fishery industry in 1965, Dr. Weinberg assembled data that indicated that approximately two-thirds of the weight of fish landed on the Canadian Atlantic coast end up as fish meal for animal feed. About one-third of this meal came from filleting waste; the other two-thirds were derived from herring. Furthermore, the Northwest Atlantic yielded substantial quantities of unexploited schooling fish such as capelin, sand lance, and argentines which Weinberg reckoned could be brought in economically if a market existed. In addition, fishing companies reported that a large proportion (25 to 40 percent) of their catch was thrown back by the fishermen because it did not meet market demand. Together with other experts, Weinberg was of the opinion that the waters of Canada's Atlantic coast, at any rate, contained sufficient quantities of low-priced fish ideally suited for an FPC industry.[7]

Dr. Weinberg also voiced the opinion that the recent contribution made by Canadian scientists to the production of FPC was not receiving the industrial attention it deserved, at least not in Canada. He pointed out that the U.S. Bureau of Commercial Fisheries, on the other hand, had selected the Canadian FPC process for further development; a pilot plant had been established by the BCF near Washington, D.C., at College Park, Maryland; extensive in-house and contract research was under way investigating the Canadian process, as well as studies in nutritive value and the development of many FPC-supplemented food products; and a 50-ton per day demonstration plant was on the drawing board.

Aware of these U.S. developments and given the existence of the basic process as well as the necessary fish resource in Canada, Weinberg felt that it was time for Canada to take advantage of this industrial potential and to proceed in devising the most suitable approach to further FPC development.

As a first step, potentially interested parties were brought together by Weinberg in the mid-1960s in Toronto to examine the raw material, processing, and marketing situation in Canada. A large engineering firm in Montreal agreed to form the nucleus of an unofficial FPC development advisory board. This group included a fishing company, a food processor, the provincial government of Quebec (which was interested in locating new industries), the Fisheries Research Board of Canada, and the Department of Industry, Trade and Commerce. A feasibility study by the engineering firm examined the possibilities and problems of scaling-up the FPC process from pilot to industrial capacity, estimated the anticipated costs of plant and finished product, and looked for a suitable plant location. Interest was aroused. In 1967, Dr. Weinberg, with other staff of the Department of Industry, Trade and Commerce and with the cooperation of the Department of Fisheries, organized the Conference on Fish Protein Concentrate in Ottawa. The Conference lasted for two days (October 24–25, 1967) and was attended by over 250 guests and speakers, including 114 from the Canadian Government, 103 from industry in Canada, the United States, and elsewhere, and several representatives from the U.S. government and the United Nations. Two dozen papers were presented at the Conference.[8]

After the Ottawa Conference Weinberg set to work on a petition to the Food and Drug Directorate (FDD) to gain approval for the manufacture and sale of FPC in Canada. This was of crucial importance since no Canadian company would build a plant without prior assurance that the product could be sold in the country. This assurance could only be given if and when the FDD had officially indicated what it considered to be acceptable, in terms of raw material and process, and, most important, in terms of product characteristics. In many respects the process of obtaining approval for the product from the FDD was similar to that undertaken by the BCF with respect to the U.S. FDA. In Canada, however, there were far fewer emotional and political hurdles to overcome: the "filth" issue was never raised, and the permission to use FPC as an additive to many foods was much easier to obtain, provided that certain food labeling conditions were met.[9]

Following extensive consultations with industry and various government agencies, it was decided that the Department of Industry, Trade and Commerce would embark on a joint submission to the FDD with the Fisheries Research Board of Canada. The petition would request permission to manufacture FPC from whole herring, whole capelin, whole hake, cod trimmings, and haddock trimmings since these species were considered to be of greatest economic importance.[10]

Through the establishment of close cooperation with the U.S. BCF, Dr. Weinberg was able to include extensive data from the Bureau in the Canadian submission. The Halifax Laboratory of the Fisheries Board produced in the pilot plant the samples required for feeding tests and chemical analyses. The results of the toxicity study and all other chemical and nutritional data were finally submitted to the Food and Drug Directorate in the fall of 1969. Approval of the product was granted in August 1970 and the regulations were confirmed by the Privy Council and were published in the *Canada Gazette* on October 14, 1970.[11]

The Food and Drug Directorate also agreed to regard the fish species mentioned as representative of their entire corresponding families, again reflecting not only a more relaxed attitude than the FDA, but also a much keener desire to assist the Canadian fishing industry. This extended the scope of the approval to about sixty-seven species of fish inhabiting the waters of Canada's Atlantic and Pacific coasts. Families that were included were the *clupeidae* (herring), the *gadidae* (cod and hake), and the *osmeridae* (smelts). The final regulations included a standard for Fish Protein Concentrate and read in part, "This standardized food product will be known as Fish Protein. The standard will serve two purposes: it will protect the consumer by spelling out all the properties essential to good nutrition. It will also be sufficiently liberal to permit the manufacturer to put out a broad spectrum of products and to develop trading standards."[12] It was expected that these regulations would encourage prospective industrial manufacturers.

PRIVATE INDUSTRY PLANS TO PRODUCE FPC (1968-1971)

The stage is now set for a discussion of the only significant industrial attempt made in Canada to produce FPC commercially, the endeavors of Cardinal Proteins, Ltd., (CPL) of Halifax, Nova Scotia, which began in November 1968 and ended tragically in late 1971.

CPL was founded by the late H. Wilson (Bill) Duval, a U.S. citizen. Duval was a self-made man of enormous ambition, energy, and single-mindedness for whom the launching of the first large-scale FPC plant in Canada was an irresistible challenge. Duval embarked upon the FPC venture convinced that, having been spectacularly successful in a previous enterprise in which he had taken great odds, he would continue to be a winner. In 1967, he met several scientists working in the Technological Laboratory of the Bureau of Commercial Fisheries in College Park on the U.S. FPC project.

The first public indication of Duval's interest in FPC appeared in an issue of *Oceanology,* dated January 4, 1968, which noted that Duval had founded a new firm called Venture Aids, Inc., in Bronxville, New York, to help finance new enterprises in science-oriented industries. *Oceanology* reported that Duval was planning a major venture in the area of fish protein concentrates.[13] Duval apparently believed that the time was right to build a large-scale commercial FPC plant in the United States and that although consumer acceptaance had not yet been proved, substantial markets would develop. He expected—naively, perhaps—that he could overcome U.S. FDA resistance to FPC and that the first FPC sales would be abroad, probably through United Nations aid programs. He predicted that once in production FPC would be significantly cheaper than dried milk.

Duval initially investigated possible plant sites on the West Coast of Florida and had already surveyed locations in New England. He planned to finance and organize the FPC venture himself and probably become its chief executive officer. He indicated to *Oceanology* that he already had a good idea of the manufacturing process he was going to select, the people he wanted to staff the operation, and the manner in which the organization should be set up.[14]

Eight months later an article in the August 3, 1968 issue of *Chemical Week* reported that the projected FPC plant would not be built in the United States but in Nova Scotia, Canada.[15] It seems likely that Duval had abandoned the idea of building his first plant in the United States not only because of the greater likelihood of obtaining financial assistance from the Canadian government but also because, having been at the FPC Conference in Ottawa in October of the previous year, he had realized that the FDD attitude toward FPC was more flexible than that of the U.S. FDA.

Duval had also retained a highly qualified and well-known engineering and consulting firm in Montreal to conduct a feasibility study concerning the location, cost, capacity, and operational characteristics of an FPC plant. The study concluded that the world "protein gap" was estimated at about 8 million tons per year, that FPC was suitable to make a major contribution toward closing the gap and that the plant's production of 20 million pounds of FPC per year with a protein content of more than 80 percent could be sold at a price of \$.35 per pound and be competitive with other animal proteins.

Because of self-imposed financial limitations, Duval had instructed the consulting firm *not* to include projections for the acquisiton of a primary fishing operation as an integral part of the new venture. For this reason it was imperative, according to the Montreal firm, that the plant be located close to a sufficiently large and diverse fish harvesting and processing center from which raw materials could be delivered at a price of \$30 per ton. Canso, at the northern tip of Nova Scotia, was therefore suggested and finally selected as the plant site since it could be located close to the \$7 million British-owned fish harvesting and processing factory, Acadia Fisheries, Ltd., which also operated a fleet of modern mid-water trawlers. Acadia had expressed willingness to supply fish trimmings and fresh fish at prices that would initially vary between \$18 and \$30 per ton depending upon prevailing market conditions. Acadia would also act as agent for the FPC plant in contracting with independent fishermen to supply the additional raw material that would be needed.

Canso's ice-free harbor would assure year-round supply of raw material. An analysis of the fish harvesting situation in the Nova Scotia area, prepared by the provincial government, indicated the reliable availability of fish which was, of course, one of the major concerns of the FPC promoters. The report estimated that since a number of undersized fish were taken incidental to the ordinary catch, on the average 30 percent of the fish caught were currently discarded and could be used in the plant. Other species, such as small hake, monkfish, and angelfish, which had no present commercial value, were also plentiful around the coasts of Nova Scotia. Furthermore, if additional processing facilities were available at Canso, Nova Scotia, midwater trawlers operating in the Gulf of St. Lawrence would stop more frequently at this port to discharge large catches of herring. Thus, according to the provincial government, the following raw materials were potentially available: between 15 and 20 thousand tons annually of edible fish hitherto discarded from Canso boats; between 7 and 10 thousand tons annually of herring; and 20 thousand tons of sand lance.

As of the fall of 1968, the raw material supply situation in the Canso area looked promising. On November 14, Duval announced the formation of Cardinal Proteins, Ltd., a private corporation owned by Venture Aids, Inc., of Bronxville, New York. Located in Canso, Nova Scotia, the multimillion dollar processing facility would produce fish protein concentrate for human consumption. With an initial capacity of 20 million pounds of FPC per year, the FPC plant was scheduled to begin operations by the end of 1969.

Duval predicted with optimism typical of most FPC endeavors that, in addition to providing a low-cost supply of nutritious high quality animal protein, the Canso facility would provide new outlets for Canadian fish products. The new market created by the protein concentrate venture was expected to have a beneficial effect on the productivity and profitability of the modern fishing fleet that existed in Nova Scotia by utilizing portions of the catch that had hitherto not been readily salable. It was also anticipated that CPL's initial production facility would serve as a prototype for similar plants to be located throughout the world for the production of vital protein supplements for underdeveloped areas where food supply was low.

This hope of creating a prototype was identical with that of the UN Children's Fund (UNICEF) in its establishment of the plant in Quintero, Chile (see chapter 8). There was, however, a difference between the UN's efforts in Chile and those of Cardinal Proteins, Ltd., in Nova Scotia in that material that was to be produced in South America was intended for use primarily in the country of origin while there was little hope that the Nova Scotia product would be used in Canada, a country with very little (or no) lack of protein. The principal beneficiaries of the Nova Scotia effort were, therefore, to be investors, industrialists, and, supposedly, the local Nova Scotia fishing population.

During 1968, Duval, as President of CPL, assembled a board of directors, made the decision to select the processing method developed in the Halifax Laboratory, and also successfully assured financial support. It was estimated that CPL would require $4,300,000 to construct and bring the FPC plant into operation. The plant and ancillaries were budgeted to cost $3,225,000; $1,075,000 was required to cover the cost of working capital, start-up expenses, and operations until break-even production was achieved. Since the Canadian authorities encouraged projects that would develop their fishing industry, generous credit facilities were available to a plant of this kind. An application for a $2,000,000 mortgage loan was made against the cost of the plant to the Industrial Development Bank of Nova Scotia with assurance from the Bank that the loan would be granted. The mortgage would be available over a 15-year period at 8 percent with amortization payments to start after three years. The project also qualified to receive a grant under the Area Development Incentives Act from the federal regional economic expansion department of about $690,000 to be used in the creation of the plant. This grant was payable in installments beginning at the start of operations although it was a bankable credit in its full amount. Finally, a line of revolving credit of up to $500,000 was to be made available by the Bank of Nova Scotia.[16]

The anticipated financial picture at the end of 1968 comprised $1,700,000 of equity funds contributed by the first U.S. investors.[17] The prospects for a mortgage of $2,000,000 and a development grant of $690,000, together with a $410,000 loan from the Bank of Nova Scotia, would provide a total amount of $4,800,000. This amount included a $500,000 contingency reserve over and above the $4,300,000 required to construct and bring the FPC plant into operation.

On April 10, 1969, Nova Scotia Trade and Industry Minister T. J. McKeough, M.D., approved a loan to CPL in the amount of $2,000,000. This loan was influenced by the fact that a grant in the amount of $690,000 to CPL had also been approved under the Federal Area Development Incentives Act. Thanks to the assistance of both the Canadian government and the provincial government the financial base for the company seemed to be assured.

In February 1969, Duval contacted the Chairman of the Fisheries Research Board urging him to start further research on raw materials other than those presently under scrutiny by the Canadian Food and Drug Directorate (FDD). Duval foresaw that fish such as skate, dogfish, sand lance, redfish, and flounder would someday be processed in his plant and wanted to start early with a submission to the FDD to obtain permission to manufacture FPC from these species.[18]

In April 1969, Duval prepared a document to interest further potential investors in the Cardinal project, detailing the selected plant site, the plant's proximity to Acadia, and the way that the raw material was going to be supplied and processed by the Halifax/BCF process into FPC. His message to potential investors also mentioned optimistically—and justifiably so—that the FDD was expected to act favorably on the petition that had been submitted by the Department of Trade and Industry, in cooperation with the Fisheries Board, for the approval of FPC prepared from whole herring, capelin and hake, and cod and haddock trimmings.

When Duval announced the formation of CPL in November 1969, he had predicted that the plant would start operations by the end of that year. This would have been an almost impossible feat under even the most ideal circumstances. As it turned out, construction work on the plant site in Canso had barely begun at the end of 1969, but, in spite of this delay, reports to the stockholders continued to be very optimistic. In January of 1970, reports were projecting the completion of machinery installation by the following May.[19]

All the planning and hard work in the world, however, could not have prevented the labor revolt which erupted within the rank and file of the Nova

Scotia fishermen in March of 1970, although it must have been evident to many observers that such an upheaval was imminent. Up to then, fishermen had not belonged to a workers' union since no legislation existed to permit them to unionize. A leader of the United Fishermen, an organization not recognized by the Canadian Labor Congress (an official organization to which all labor unions in Canada must belong to be accredited), had arrived from the Canadian West Coast where he had organized the fisheries labor force. He now tried to persuade workers in the Nova Scotia fishing industry to join his union in order to bargain for better working conditions.

The attempt did not fall on deaf ears. The fishermen felt that they had been working without proper health insurance, guaranteed vacations, pension fund, and other benefits. In general, they felt their working conditions to be very difficult and unfair: enormous perils inherent in fishing operations on the high seas, great hazards involved in marketing their produce, income and livelihood uncertainties to which they are subjected, and the traditionally low social status to which members of the industry have been relegated for as long as anyone could remember.

Fishing vessel crews, as well as workers processing fish at Acadia and other plants in the province, went on strike and, as a consequence, other trades such as pipe fitters and welders, decided to conduct a work slowdown in a gesture of sympathy and support for the fishermen. Chaos ensued which was immensely costly to the province and to all the industries situated there. Several large firms, including a Gulf Oil refinery, a Michelin Tire plant, a Heavy Water plant, and of course CPL, were all in the process of plant construction.

Erection of equipment at CPL proceeded more or less on schedule into late spring. Welders and pipe fitters were desperately needed in June but were almost impossible to hire. The months of July, August, and September of 1970 cost Cardinal several hundred thousand dollars since, in order to make any headway at all with construction, Cardinal was forced to pay exorbitant wages. In spite of these expenditures very little, if any, progress was made in CPL's construction plans. In the fall it became clear that the company needed more money very quickly to offset the overrun of $750,000 that the company had suffered because of the workers' refusal to do more than token work while demanding full pay.

An appeal for help was made to the provincial government. However, Nova Scotia was in the midst of preparing for an election which only added to the confusion, and, in October 1970, the Nova Scotia government changed. Yet, despite all the problems the new administration was having with the labor situation, government officials agreed in principle to guarantee a new bank loan that would permit Cardinal to survive.

A day or so after government officials had made their decision to help CPL, Duval suddenly and without consultation decided to refuse the loan;

apparently, he did not want to take on any additional financial burdens and was also unwilling to raise more money on his own.

Although the CPL board of directors had expressed their confidence in Duval, the CPL President seemed a sick and broken man and disappeared from the scene for a month, presumably to rest. On January 6, 1971, he committed suicide at his home in Bronxville, New York. Tireless and courageous efforts were made by one of the other directors, who had been elected Acting President, to save the situation at Canso. In the fall of 1971, however, the firm went into receivership, a month after CPL's fish processing neighbor, Acadia, Ltd., had declared bankruptcy. The Nova Scotia government and others tried hard to interest a number of industrial buyers (including, we believe, the Astra Group of Sweden) to jump into the breach and take over the plant to continue the FPC project; they were, however, unsuccessful, and the Cardinal Proteins plant was finally sold piecemeal.

It is useful, at this juncture, to look back over the Canadian FPC experience. In retrospect certain events and conditions in Canada clearly assisted the development of FPC significantly, while others created important obstacles to the initiation of an FPC industry, some of which proved to be unsurmountable.

Among the events that were clearly supportive of FPC developments in Canada were, first and foremost, the FPC-related discoveries and scientific advances made in the Halifax Laboratory. There was also the development and active government support (over several years) of the basic idea of utilizing fish for the preparation of a dietary protein supplement. Other positive factors included organization by the Canadian government of the FPC Conference in Ottawa, the financial support proferred by federal and provincial authorities to Cardinal Proteins, Ltd., and the cooperative attitude of the Food and Drug Directorate of Canada in approving the application by the Department of Commerce, Trade and Industry to manufacture and sell FPC in Canada.[20] The distribution by the Halifax Laboratory of FPC-related information and FPC samples to other scientists all over the world also supported FPC developments elsewhere.

The general atmosphere of public opinion that prevailed in Canada during the years under review was also supportive. There was no overt public or industrial opposition (as occurred in the United States) to the idea of FPC as a food additive, and the Canadian government's policy to assist and expand the Canadian fishing industry was well known and well supported.

The FPC-related events that both helped and hindered the course of FPC developments in Canada were represented by the actions of H. W. Duval. On the one hand, the single-minded devotion and relentless energy of CPL's founder and president brought him closer than anyone else in Canada to the goal of establishing a modern FPC industry; on the other hand, Duval bears part of the responsibility for CPL's demise. The mainspring in his personality that propelled him with great momentum toward success also made him inflexible. It prevented him from accepting and reacting constructively to evidence that his predictions about CPL's financial prospects, raw material supply, and market opportunities were all oversimplified. It incapacitated him in the face of prolonged difficulties of various kinds and made it inconceivable for him to face possible defeat. Why did Duval not invest more of his own money in CPL? Why did he not accept the offer of help from the Nova Scotia provincial government? These and other questions surrounding the failure of Cardinal Proteins, Ltd., and the sad demise of its founder will remain unanswered.

The events that created outright obstacles to the development of FPC in Canada include, of course, the labor dispute in which the Nova Scotia fishermen, the Canadian Labor Union, the federal and provincial Governments, and the Nova Scotia industries were deeply involved and in which many people and organizations suffered grievous losses. What is significant here is not so much that the occurrence of the revolt of the Nova Scotia fishermen during 1970 was a most unfortunate coincidence in so far as Canadian FPC development was concerned, but that the conflict was by no means a fluke or an isolated event, but only one manifestation of the problems afflicting the Canadian fishing industry at that time and should have been anticipated since the conflict's origins were of very long standing. The conflict offered another basic element for consideration in assessing the feasibility and likely success of technological innovations, namely, the close and important tie between the raw material for the manufacture of a food and the social, political, and economic context within which this raw material is produced and harvested.

The failure to establish an ongoing FPC industry, however, should in no way cloud the real contribution of the Canadian government and the Halifax Laboratory to the development of FPC through the government's willingness to support the research and the laboratory's preparedness to share the results of its work.

Chapter 7

FPC in Morocco: Conflicting Aid Objectives

One of the major international efforts to establish an FPC industry in a developing country took place in Morocco. In one way or another this effort occupied the attention of officials of several national governments, namely, Morocco, the United States, and France, various United Nations agencies such as the UN Industrial Development Organization (UNIDO), the Food and Agriculture Organization (FAO), the Protein Advisory Group (PAG), and the UN Development Program (UNDP), not to mention a variety of private entrepreneurs and at least one charitable organization for fifteen years or more. The intangible results may be difficult to assess at a distance, but the tangible results are clear: by the end of the UNDP program of financial investment and technical assistance in 1973, there had been no commercial production of FPC for human use, nor is there any to date. The history of FPC in Morocco provides an opportunity to examine the roles of the UN agencies, the U.S. Agency for International Development (AID), and the Government of Morocco in a multipartite effort that reflected the conflicting interests of the various institutions and individuals involved.

BACKGROUND

Although Morocco has long had a well-established fishing industry, in common with its neighboring countries on the north and west coasts of Africa, its per capita consumption of fish is low. In 1959, total domestic landings were 144,000 tons.[1] FAO statistics for 1973 (the year the UNDP project terminated) give another indication of domestic consumption. Of the total domestic landings of 397,000 tons, 189,000 tons went to direct human consumption, 120,000 tons of which were exported, leaving only 69,000 tons for domestic consumption, which is the equivalent of 4.2 kilograms per year per person. (In fact, the major part of this domestic fish consumption is accounted for by the foreign population in the urban areas, and only a minor part by the coastal population; three-fourths of the Moroccan population lives in rural areas, is poor, and traditionally prefers meat over fish.) Most of the 208,000 tons for nonhuman consumption went into exported fish meal, with only 8000 tons being retained for domestic use.[2]

EARLY INTEREST IN FPC: THE INITIATION PERIOD (1958-1962)

The Moroccan FPC story begins in 1958, at which time a fish meal industry already existed, based on the Mediterranean sardine, the dominant species group harvested.[3] Experiments were begun to develop an FPC process in a pilot plant constructed by the Société de l'Union d'Azote in a factory in Safi owned by the Société des Sous-Produits de la Pêche, a fish meal and fish oil company. The project was undertaken by Azote-Union (as it became known) with the assistance and at the urging of Pierre Tord, a French chemical engineer, who was convinced that a substantial market existed for FPC manufactured in Morocco. It is not clear from the evidence available if M. Tord's projections depended primarily on a domestic or an export market.

UNICEF and the UN Food and Agriculture Organization were at this time becoming interested in the potential of FPC as a protein supplement, UNICEF as a consequence of its concern with feeding children and FAO as a result of its interest in fisheries and food. FAO became interested in the Moroccan project by 1960 and provided technical cooperation in the form of consultation and assistance in performing chemical analyses.[4]

The pilot plant in Safi was used to develop a two-solvent process for defatting fish meal made by cooking, pressing, and drying whole sardines (*Sardina pilchardus*). The meal was extracted first with alcohol, then with hexane, and then dried under vacuum. The dried, defatted meal was ground, screened, and finally packaged. The Safi plant eventually produced about 150 kilograms of FPC with a maximum production rate of 600 kilograms per day.[5]

By July 1960, the plant was producing at the rate of 10 tons per month, and its product had been tested for nutritive value and analyzed in several European and U.S. laboratories, thanks to the cooperation of FAO. The results of these tests were somewhat ambiguous, at least as they were reported to the Protein Advisory Group (PAG) of the United Nations in 1960. On the one hand, the report pointed out that "samples subjected to laboratory examination in the Netherlands of the early production showed high bacteriological counts" while, on the other hand, "the Nestlé Laboratories in Switzerland have made a thorough examination of a sample of the flour and found it completely satisfactory."[6] The apparent contradiction may simply have resulted from Nestlé's having received a sample from a later production run, after processing improvements had been made. The report concluded by stating that the plant's output was being used locally.

A further illustration of the confusion that seems to permeate the record of this project is the report presented at an international conference at Abidjan in 1964 in which the outcome of those analytical results was described as follows: "The collected analytical results were examined by the Protein Advisory Group at its meeting of 3 June 1961 in Washington, at which time it concluded finally that the [fish] flour produced at Safi met established standards and could be used for human feeding to supplement and balance protein-deficient diets."[7] The record shows, however, that the PAG met in New York in 1961 and did not discuss the Moroccan FPC project at all. In fact, there is no record that the PAG ever specifically "concluded" that the product of the process developed at Safi met the standards required for human consumption and "could be used to supplement and balance protein-deficient diets."

In any event, the quality of the Safi product eventually improved sufficiently[8] that, on the recommendation of FAO, the Ministry of Public Health encouraged the use of FPC in a variety of child-feeding programs.[9] This effort took place under the direction of Dr. François Rémy, Head of Preventive Medicine, Ministry of Public Health, with the technical cooperation of FAO. Acceptability studies were conducted in hospitals, homes for the aged, orphanages, schools, and maternal and child health care (MCH) centers with apparent success.[10] Much of this testing was done under the aegis of the Alliance Isráelite, a Jewish welfare organization in North Africa and the Mediterranean area, which provided lunches supplemented with FPC to approximately 50,000 children in Jewish schools.[11]

A COMMERCIAL ENTERPRISE FOR FPC PRODUCTION: THE NATIONAL PERIOD (1962–1966)

The results of these feeding studies were sufficiently encouraging that in 1962 a joint agreement between Azote-Union, Pierre Tord, and the Bureau d'Etudes et des Participations Industrielles (BEPI, an agency of the Moroccan government concerned with participation in industrial development) created SONAFAP, the Société Nationale de Farine Alimentaire de Poisson, a 93 percent government-owned enterprise.[12] Named as Director General of the enterprise was Abdellah Jaidi, brother of Mohammed Jaidi, Minister of Commerce, Industry, Mines, and Merchant Marine.

SONAFAP was capitalized at two million dirhams (US $400,000)[13] for the purpose of construction and operation of an FPC plant at Agadir designed to process 50 tons of sardines per day and produce between 1300 and 1400 tons of FPC during the seven-month fishing season. Additional funds were invested in the project for, as of December 1966, the total capital investment was reported at $960,000, including development, land, plant equipment, office facilities, and trucks.[14] (Some of this funding was devoted to expanding production capacity since, by 1970, the plant was reported to be capable of processing 150 tons of sardines daily, making fish meal.)[15] By 1965, the new plant at Agadir was in operation and produced 32 tons of FPC in 1965 and 145 tons in 1966.[16] During the initial operation of the plant in 1965 M. Tord was on hand to assist; however, his technical connection with the project seems to have ceased in 1965. Technical assistance during the 1966 operating season was provided by Assistance Technique Française (a French government agency), which supported the services of two engineers.[17] The FPC produced at Agadir was principally intended for use in government institutions, but due to unacceptable odor and taste, did not come up to the quality of the FPC made previously at Safi. Consequently, the Agadir plant was shut down in September 1966 pending a resolution of its difficulties.[18]

THE UNIDO PROJECT AT AGADIR: THE INTERNATIONAL PERIOD (1967–1973)

In an effort to salvage the Agadir operation, a UN mission, consisting of a representative of UNIDO, two FAO representatives, and a consulting engineer from the United States, visited the plant early in 1967 at the request of the Government of Morocco. The mission was charged with examining the facilities and determining if an acceptable FPC could indeed be manufactured, taking into account the raw material, the process, and the equipment. The result of that visit was a proposal to UNIDO which used it as the basis for a proposal to UNDP for a four-phase approach toward resuscitation of the SONAFAP plant and establishment of an FPC industry in Morocco.[19] Phase I called for a conclusive demonstration that an acceptable FPC could be made from the sardines used as raw material by SONAFAP. Phase II involved experiments, at the plant, with alcohol extraction of fish meal, extraction of undried presscake (the material remaining after the fish is cooked and pressed), and, finally, extraction of raw fish. These experiments were intended to provide the engineering data necessary to redesign the Agadir plant for production of FPC of acceptable quality.

Phase III of the UN mission's recommendations called for a market study and a study of possible uses of FPC in Morocco, to be conducted by AID, which, at the end of 1966, had been designated in the United States as the "lead agency" for the funding and administration of the Food from the Sea Program.[20] It was recommended that approximately $20,000 be allocated for this study. The fourth and final phase of the UN plan was the revision and reconstruction of the plant along the lines indicated in Phase II.

In considering the UNIDO proposal, the United Nations Development Program, which was responsible for funding special projects of UN agencies, made its commitment to fund the project contingent on AID's willingness to carry out the proposed FPC market study.[21] AID eventually did indeed agree to assume this responsibility and earmarked $180,000 for the study in July 1969.[22]

Phase I of these recommendations was carried out with the cooperation of FAO, UNIDO, and the U.S. BCF. BCF's College Park Laboratory, College Park, Maryland, undertook to produce FPC from Mediterranean fish supplies through the efforts of the UN. The BCF report on that experiment concluded that "NCFPC's [National Center for Fish Protein Concentrate] investigation on production of fish protein concentrate (FPC) by isopropyl alcohol extraction of Mediterranean sardines indicates that the product meets all present U.S. Food and Drug Administration requirements on chemical composition and nutritive value. The products were made both by laboratory procedures . . . and by . . . procedures that approximate commercial production methods. It is particularly significant that, in agreement with results obtained for FPC made from other species of fatty fish in this laboratory, a satisfactory FPC can be produced from Mediterranean sardines by this prototype commercial process."[23]

While this work was in progress, AID was exploring possibilities for FPC demonstration projects in developing countries in Asia, Africa, and Latin America. As part of its Food from the Sea Program, AID's Office of the War on Hunger (responsible for the Program) had made preliminary surveys of selected countries in Latin America and Asia (September and October 1967 respectively), and in January and February 1968 a survey team visited Morocco, Ghana, and Uganda "to identify, if possible, a suitable country in which to mount an FPC demonstration program." AID also charged the survey team with evaluating "the future possibilities of the existing plant at Agadir. . . ."[24]

This team concluded that Phase II of the UN mission's recommendations should indeed be carried out, but it disagreed on the necessity for the third phase as planned. The report pointed out that nutritional and acceptability studies had already been made, with mixed results. These studies, carried out by SONAFAP with the collaboration of the Ministry of Public Health and the Alliance Israélite, had "already demonstrated that the material (SONAFAP FPC) is nutritional [sic] and nontoxic, could be fed to a captive audience, but did not find ready acceptance in family diets."[25]

This opinion was consistent with the previously cited experience of the Alliance Israélite and with conclusions drawn by FAO in earlier appraisals of FPC as a food commodity in Morocco. In a report on this subject prepared for the PAG at its July 1965 meeting in Rome, FAO's G. D. Kapsiotis had pointed out that "the preparation of baby food in maternal and child health

(MCH) centers in hospitals does not pose problems if those who are responsible for running them understand the value of FPC." He also noted that FPC produced at Safi had been used for some years by biscuit manufacturers in Morocco as part of the acceptability studies already referred to. But because of the "lengthy and difficult education campaign . . . necessary to teach mothers in low-income groups" to mix FPC with the traditional constituents of their children's foods and the additional cost of the FPC itself, Kapsiotis concluded that "it is not feasible . . . to depend upon the housewife for the supplementation of food with FPC" if it costs more than the ingredients it might replace.[26] Furthermore, the economic barriers to the development of a significant market in Morocco for FPC as an ingredient in manufactured foods were illustrated by Kapsiotis with respect to FPC-enriched bread, a product that would be one of the first logical candidates for such commercial use. Since the price of regular bread was controlled by the government, bakeries could not hope to incorporate FPC if its cost were more than that of the flour—unless, of course, it was subsidized by the government. This left the so-called special breads, whose price was not controlled, as a possible vehicle, but consumers of these breads were not those in need of nutritional supplementation. Couscous was another possible vehicle for FPC since trials had shown that a completely acceptable FPC-supplemented dish could be made.[27] Most couscous, however, is homemade and, as is typical with traditional foods, its composition and preparation are not easily tampered with. Nor would commercial baby foods have been a sensible vehicle because the small amount sold was mostly imported and used only by upper-income groups.

In 1965, the cost figures at Agadir were such that the projected price of FPC would have had to be reduced by a factor of 4.3 (from the equivalent of \$.60 per kilogram to \$.16 per kilogram) before the Office Chérifien Interprofessionel des Céréales, which controlled the manufacture and the price of wheat flour in Morocco, could consent to its use as a wheat flour supplement.[28] Thus, the situation in Morocco was even worse than perceived at the time, for the \$.60 price projected for FPC was probably low by a factor of two.

In 1968 the AID survey team recommended strongly against the use of SONAFAP FPC for additional studies in Morocco until and unless the quality of the product could be improved sufficiently to be comparable with the FPC produced by the BCF process. Instead, they urged that the SONAFAP plant be used to produce a feed-grade FPC to be sold as a high-protein supplement for animal rations.

Although there are some reports that the AID mission in Morocco was convinced of the necessity for the market study,[29] the mission director at that time, Philip Birnbaum, is unequivocal in stating that he opposed the project from the start and that he was "dumbstruck" at the thought of the SONAFAP plant producing a food product such as FPC, given its location, facilitics, and the experience and capabilities of its personnel.[30] Nevertheless, in spite of this stand and the report of the AID survey team, AID inexplicably moved ahead the following year with a commitment to undertake the market study using SONAFAP FPC.

Meanwhile, overlapping the experiments at College Park, Phase II experiments were being carried out at the SONAFAP plant. Thus, for a while a heavy flow of letters and cables passed between Agadir and College Park. Samples from various stages of processing at Agadir were sent to College Park for analysis, the results of which were needed by the engineer at Agadir to obtain the data needed for Phase IV, that is, the modification of the plant for reliable production of acceptable FPC.[31]

Following completion of the Phase I experiments in June 1969 and the Phase II experiments a few months later, a joint UNIDO/FAO expert group meeting was held in Rabat, December 8–12, 1969. This meeting brought together many of the people who had been concerned with FPC development over the years. One of the two stated objectives of the meeting was a discussion of "the development of a commercial enterprise to produce FPC in Morocco, taking into account the efforts made so far."[32] The other was the general discussion of commercial production and distribution of FPC-supplemented products. Seventeen formal presentations were made, covering the history of FPC, processing problems, and utilization. Only two of the papers presented, however, dealt with the Moroccan endeavor. The conference concluded with a series of recommendations. These included the recommendation that technical assistance be provided SONAFAP for:

a. changes needed in the SONAFAP plant to produce a good quality, low-fat fish meal as well as ordinary fish meal;
b. production of FPC (for human consumption) for marketing purposes, and development of FPC food products;
c. problems connected with pilot production of FPC foods in Morocco; problems of technical and economic management and training of personnel.[33]

Of the two conference papers directed to the Moroccan effort one discussed the engineering problems of the SONAFAP plant.[34] The other, given by an AID representative, described USAID goals in Morocco as being "to determine the feasibility of establishing an economically viable fish-protein concentrate industry in Morocco and to develop a method of introducing FPC and FPC-fortified products into the Moroccan food system; and to develop a marketing and distribution plan for feed-grade FPC produced by the SONAFAP plant at Agadir." The paper went on to state that AID, in consultation with UNIDO and FAO, hoped to achieve these goals by

determining the economic potentiality of converting sardines into FPC and the demand required to justify the operation of an industry in Morocco;

establishing the cost/protein value of utilizing FPC as opposed to the cost/

protein value of alternative proteins or mixes of proteins available or potentially available for use in Morocco;

developing new food formulae utilizing FPC, testing the acceptability of all FPC-fortified products through target groups, and setting up a system whereby these products can reach the target groups; and

analyzing the economic practicality of producing feed-grade FPC at SONAFAP, its marketing and distribution.[35]

Two points are immediately apparent from the two papers specifically concerned with the Moroccan FPC endeavor. The first is that the only mention of the four-hpase recommendation of the 1967 UN mission (which had provided the stimulus for holding the conference) was made by the engineer who was part of that mission and who presented the paper discussing the SONAFAP plant. The second is that the AID presentation, while it included the comments noted, did not refer once to the "responsibility" for carrying out Phase III that AID had already assumed by its funding obligations of funds.[36] Indeed, AID appeared to be hedging its concern for "introducing FPC and FPC-fortified products into the Moroccan food system" with its additional goal of developing a plan to market feed-grade FPC from SONAFAP, consistent with the recommendation of its 1968 survey team.

Thus, in spite of its previous commitment to UNDP it seems safe to conclude that AID/Washington was already moving away from the idea of producing FPC for human consumption in Morocco toward that of manufacturing feed-grade FPC for the livestock market in Europe. Production of FPC for human consumption would have necessitated costly changes in equipment and improvements in processing and handling raw material which the manufacture of a feed-grade material, on the other hand, would not require. Furthermore, a feed-grade FPC project would eliminate the need for a difficult and perhaps costly market survey in Morocco.

More important in AID's change of direction, though, was the shift in policy of the new administration in Washington. With Richard M. Nixon's election to the presidency and the attendant departure of Vice President Hubert H. Humphrey from the White House, the U.S. FPC program had lost its most important political advocate. The controversty over FPC production costs[37] and the decline, off U.S. shores, of previously large stocks of unutilized fish also contributed to the lessening of official enthusiasm for FPC. In addition, the overwhelming concern of the new administration with the war in Vietnam and domestic opposition to the war relegated programs such as FPC to levels of minimum importance. The effects of this shift became evident in the tightness of the BCF's research budget for FPC in FY '72 (that is, beginning July 1971).

Another aspect of the recommendations of the Rabat conference is particularly significant in evaluating the sincerity with which all parties regarded the necessity for the proposed market study. Specifically referring to "feasibility and marketing," the conference not only recommended that "market and feasibility studies . . . necessary to establish an FPC industry in Morocco" be carried out, but it took pains to point out subsequently that the studies should also be aimed at determining the "economic validity of producing . . . low-fat fish meal for animal feed.[38] Finally, the recommendations emphasized that the rehabilitiation of the SONAFAP plant should be consistent with—

a. the need for an adequate supply of good quality FPC for market studies and other immediate demands;
b. increasing the production of fish meal; and
c. the plan to produce low-fat fish meal.[39]

It is important to note that, in reality, two separate studies were being recommended, since the only thing that FPC and low-fat fish meal have in common as products is the raw material—and even that is open to question in terms of the different standards of fish quality used for human food and for animal feed. Under the circumstances obtaining in Morocco, fish meal was strictly an export product with a known market. The major question was what premium the feed manufacturers might be willing to pay for a low-fat product, compared with the cost of producing that product. A market study certainly needed to be conducted to justify investment in such an enterprise. On the other hand, FPC for human consumption was a product that might be used for either domestic sale or for export. Without an established market, however, a separate market study was needed in either case.

The conference recommendations clearly ignored the one clear technological economic lesson learned from experience with FPC and fish meal manufacture, namely, that the sanitary manufacture of nutritious, wholesome FPC is incompatible with the manufacture of conventional fish meal in the same plant. The materials of construction, processing equipment, materials handling, and packaging procedures necessary to assure a high-quality food product such as FPC simply cannot be economically justified in a fish meal operation. A fish meal plant using off-grade material might make technical and economic sense it were associated with an FPC plant on the same or a neighboring site, perhaps, but not if the two were combined in one plant. Therefore, it is difficult to believe that the conferees at Rabat had serious expectations of FPC production by SONAFAP in view of their recurring emphasis on "feed-grade FPC," another name for low-fat fish meal.

The bureaucratic stage at this point was full of actors with conflicting roles. In summary, an international conference was called by UNIDO in December 1969 to discuss the development of an FPC industry based on the

SONAFAP plant. UNIDO's interest stemmed from a 1967 mission that recommended that AID do a market study for FPC and FPC-fortified food products. The project would receive UNDP support only if AID agreed to do the study. AID, whose supposed interest is in food, not feed, was the only agency in a position to support such a study, and it was already experiencing internal dissension about its usefulness. The conferees recommended a course of action that, when closely examined, clearly said "fish meal" and not "FPC." If the message was received, AID would not do the study, UNIDO would not receive UNDP approval, and SONAFAP would not receive its funds. At the same time, all signs pointed to negative results from such a study if it were done, in terms of both a domestic and an export FPC market. Thus, if the study were performed before the UNDP reconstruction project was started, much, if not all, of the funding would cease.

In either case SONAFAP would lose. If AID unilaterally reneged on its commitment, its relations with both UNIDO and Morocco would suffer. If it went through with the study, it would probably be wasting its money and effort and would perhaps suffer repercussions within the Executive branch in a period of dwindling interest in food from the sea. Clearly, it was in AID's best interest to abandon the study, but under circumstances that would not endanger its relations with UNIDO and Morocco. It was in SONAFAP's best interest to have AID's commitment maintained until the UNDP project was well under way, but then to have the study cancelled. It was in UNIDO's best interest, as the enthusiastic originator of this project,[40] to have it funded by UNDP but not to have the market study performed after all, lest it prove the original concept ill-conceived. How these conflicting interests were resolved becomes clear from postconference moves.

After the Rabat conference the original UNIDO plans moved forward and negotiations for undertaking the UNDP-financed reconstruction of the SONAFAP plant (Phase IV) continued. On December 17, 1970, the final signature was affixed to the agreement when Gerdt Lovold of UNIDO was named Project Director and the official commencement date was set for April 1, 1971. (The project was originally scheduled to be completed in two years but was eventually extended for three months, to June 30, 1973.) Designated as the cooperating government agency was the Moroccan Ministry of Commerce, Industry, Mines, and Merchant Marine, whose Minister at the time was Mohammed Jaidi, brother of the Director-General of SONAFAP. The agreement also designated UNIDO as the agency charged with execution of the project and FAO as "Associated" with the execution of the project.[41] The Government of Morocco agreed to contribute 476,900 dirhams ($95,380) in goods and services and 87,000 dirhams ($17,400) in local currency while the UNDP's contribution was fixed at $289,900. The agreement was signed, however, with no formal contract made for the AID market study.

AFTERMATH

The final SONAFAP project report, dated 1974, describes the goals of the project as follows:

> The aim of the project was to furnish assistance to the SONAFAP plant over a period of two years and, if the results of the market study to have been performed by USAID were positive, to assist the government in determining if production of fish protein concentrate destined for human consumption and animal feeding could be developed and sustained on a profitable basis.

The report goes on to state:

> The assistance to be furnished the SONAFAP plant was to be concerned especially with the following activities:
>
> quality control of production of FPC for human consumption and animal feeding;
>
> improvement of operation of plant equipment;
>
> improvement of fish handling at sea and ashore as well as degradation and storage of FPC;
>
> technical studies aimed at improvement of quality, increasing production capacity, and reducing production costs;
>
> commercialization of defatted fish flour [farine de poisson améliorée—FPA];
>
> preparation of experimental FPC-fortified foods with acceptable flavor, odor, and color;
>
> establishment of a feasibility study aimed at determining the investment needed to launch a profitable FPC industry in Morocco, based on the recommendations of the USAID market study.[42]

The reliance on the AID supported market study as the basis for the UNDP project is emphasized throughout this report, in spite of the fact that the project was begun with no agreement for such a study between AID and the Moroccan government.

After outlining the goals of the project and the nature of the assistance to be rendered SONAFAP, the report notes that the all-important market study was cancelled at the request of the Government of Morocco and points out the serious change in plans necessitated by this turn of events. Curiously, no mention is made of the official date of cancellation, June 29, 1972,[43] yet the report describes the efforts of the UNDP and the Moroccan government to reach agreement on a *revised* program as having started six months earlier, in January 1972.

In any event, an agreement on the revised program was eventually reached on April 15, 1973, some fifteen days after the project was originally scheduled to terminate. Not surprisingly, one of the major bones of contention was the amount of UNDP funding that would be retained in the revised program, and it was not until UNDP agreed to retain its commitment to spend the full amount of the original $289,900 that the Government of Morocco agreed. The project as revised concerned itself principally with improvement and replacement of equipment, reduction of steam losses, increasing extraction capacity for production of FPC, evaluation of the sardine fishery and trawler requirements to supply the plant, economic analysis of FPC production, reduction in amount of isopropyl alcohol required for extraction, and improvement in fish handling.

Finally the revised plan also included a seven-week study by an FAO nutritionist who reached the following conclusions:

SONAFAP FPC constitutes a remarkable source of animal protein and offers the possibility of improving the diets of the majority of the population of Morocco.

Consumer reaction to SONAFAP FPC is negative and, under the circumstances, no commercial or industrial enterprise will take the risk of trying to commercialize FPC either as a simple food or as an ingredient in food mixtures.

To remedy the situation would require an intense publicity and educational campaign on nutrition, via radio, television, etc., associated with a propaganda effort aimed at increasing consumption of fish.[44]

As a consequence of the favorable fish meal market and the increase in price of other protein concentrates used as mild replacers in animal feeds in 1973, the SONAFAP plant was able to begin exporting FPA for piglet starter rations in Europe.

The conclusions reached by UNDP after the termination of the project on June 30, 1973, are stated in terms of the recommendations of the group of experts involved. They concluded that, in the absence of an FPC market, the most profitable course for the SONAFAP plant would be to make the best use possible of its fish meal manufacturing equipment and produce FPA. This could be accomplished through the following steps: (1) immediate installation of a centrifuge for oil recovery (it was calculated that oil recovery would have provided an income of 390,571 dirhams (roughly $97,650) in 1972); (2) overhaul of processing equipment and stocking of spare parts; (3) installation of a new boiler and water treatment system; (5) installation of a continuous extractor complete with solvent recovery system; (6) chartering one or two 150-ton trawlers for exclusive use of the plant.

The Agadir plant is still operating, exporting FPA for piglet starter rations, but none of the changes recommended by the UNDP has been made. One of the consequences of the effort, however, for which the UNDP undoubtedly can take much deserved credit, is that three other FPA plants are now operating in Morocco.[45]

It is useful here to consider the machinations whose end results were that the market study was not done: USAID was not financially involved; UNDP funded the project despite its prime condition for funding not having been met; and the SONAFAP plant was revived for the production of upgraded fish meal, the purpose its Moroccan supporters had envisioned since 1968.

It was pointed out earlier that the agreement between the UNDP and the Government of Morocco was signed in December 1970 with no formal agreement for the market study having been made. In its July–September 1971 quarterly report the UN mission in Rabat speaks of discussions in Washington to "revive a planned market study" but states that the local USAID still had no instructions on the matter.[46] The UNDP continued to press AID and on February 21, 1972, the UN Resident Representative in Rabat reported that the USAID mission in Rabat was "fairly certain" that the study would be made, although no word had been received from Washington.[47] A week later the UNDP wrote to AID to state that it was essential that the market study be carried out as quickly as possible.[48]

By March, the UNDP was informed that USAID had officially decided to move ahead,[49] in May the UNDP Project Manager was "await[ing] the green light to produce FPC" for the AID market study,[50] and on June 20 the Resident Representative reported, after meeting with an official of USAID, that first the mission was still waiting for the Government of Morocco to sign the Project Agreement, and second, even if this did not take place, AID was willing to hold the funds for the next fiscal year.[51] On June 29, 1972, however, the UN field office in Rabat reported that with respect to the "problem of the market study," the Government of Morocco had asked that AID not do the market study "during this fiscal year" at the oral request of Mohamed Belkhayet, Secretary of State and Economic Affairs.[52] Although the New York office of UNDP had been informed of this turn of events shortly before, by telephone, this seems to be the first formal notification that the market study was being cancelled—and it was an internal UN document. This is particularly curious in that the USAID mission in Rabat received notice of the government's change of heart two months earlier on April 25, during a discussion with the Secretary-General of BEPI.[53]

The reasons given by the Government of Morocco for cancelling its request for the AID market study seemed to change according to the agency to which they were given. The UN field office reported Mr. Belkhayet's argument that the entire management of SONAFAP needed to be revamped due to its ineffectiveness but, in view of the important position of the brother of SONAFAP's Director-General, Mr. Belkhayet hesitated to take drastic measures against him and was looking for an "elegant way out."[54] The cancellation of the market study provided just such an option.

USAID, on the other hand, was informed that "revised government procedures" had resulted in BEPI's having taken over 95 percent of the Moroccan financial participation in SONAFAP early in 1972 and that SONAFAP "had completely abandoned experiments leading to the production of FPC for human consumption and did not intend to revive the project." Furthermore, Benhayoun, the BEPI Secretary-General, stated that SONAFAP was "concentrating henceforth on production of upgraded fish meal (UFM) for which there was an excellent and immediately available market."[55]

As a consequence of the government's request, AID was spared the trouble of officially reneging on its commitment to UNIDO to undertake the FPC market study, though, as we have mentioned, AID had lost interest in the study long before the inception of the UNDP project. In September 1969, AID's Bureau of Technical Assistance had come into existence, and in the course of taking stock of ongoing programs, the Bureau's Director, Joel Bernstein, expressed some doubts about the value of the FPC market study.[56] As a result, the offices immediately concerned were asked to comment on the repercussions the agency could expect if it were to drop its financial support for the study. In a memorandum to the Associate Assistant Administrator for Technical Assistance, the Africa Bureau responded in February 1970, pointing out that such a course of action would be "highly undesirable" vis-á-vis the effects on the UNDP project and noting that both the Africa Bureau and the AID mission in Rabat strongly supported the study. However, the memorandum pointed out that two issues were still to be resolved before an agreement could be reached with the Moroccan government. These were (1) the availability and amount of local currency to be supplied by the government, and (2) the government's commitment to purchase the product of the future FPC plant.[57]

At one level, then, AID was acting as if the Project Agreement for the market study would be signed before the end of FY '71; the Request for Proposals had been sent out, based on the Project Paper initially prepared in 1970,[58] and proposals were being evaluated up to the date of cancellation.[59]

It seems clear, however, that by the beginning of FY '70, AID was skeptical at the administrator's level not only about the wisdom of the market study in Morocco but also about any further activity by the agency with FPC. The bureaucratic decisionmaking process within the agency was such, however, that no decision was made. People at the lower levels of the agency went about their business very conscientiously, preparing for a study that was destined never to occur. Finally, the Government of Morocco's decision left AID free of the burden of taking any official action.

The Government of Morocco—or at least those officials with a personal or institutional stake in the financial success of the SONAFAP enterprise—must have decided as early as 1968, though for reasons different from those of AID, that the market study was not desirable. The Government was cognizant of the negative attitude the AID survey team had expressed in that year about the market study and of the team's conclusion that Moroccan-produced FPC would not be able to compete as an export product on the world market. In

FY 1970, prior to the signing of the contract for funding with UNDP, the Government of Morocco had refused to sign a Project Agreement for the AID market study, ostensibly because "(1) The UN had not presented the UN Plan of Operation, and (2) the FPC that AID had planned to use in the market study was not acceptable."[60] Both these reasons were simply a way to stall for time, for in fact the delay in presentation of the UN Plan of Operation stemmed from the central conflict between UNIDO's original proposal to provide large-scale capital assistance to the SONAFAP plant[61] and the UNDP's position "that plant modifications should not take place until a market feasibility study (AID's contribution) was completed."[62] Objections such as those raised by the government were made in order to hold out for a Plan of Operation that would in fact allow assistance to be given to SONAFAP prior to the beginning of a market study. Apparently UNDP became convinced that such a plan was feasible since the contract was signed without a definite agreement as to the start-up time or duration of the market study.

It is difficult to draw any conclusion other than that the Moroccan officials knew all along just where they were headed—toward a UNDP investment of several hundred thousand dollars in funds and technical assistance to establish a profitable business in manufacture of upgraded fish meal. Because of the conflicting inclinations of the other parties involved, and most important of UNDP, this goal had to be disguised as an FPC project that, by some combination of events, would unfortunately never materialize.

The clues to the eventual outcome of this project were there for all to see, but, for its own reasons, each of the parties directly involved chose not to see them. If any one institution was victimized, it was UNDP, for it depended on the advice of a partisan agency, UNIDO, which had initiated the SONAFAP project and wanted to see it continue. However, with competent personnel of its own—and the willingness to say no to a member country (which UNIDO also lacked)—UNDP could have avoided this situation.

In conclusion, it is instructive to note the comment of the UN Resident Representative in Rabat, in reporting on the cancellation of the market:

> The more I think about it, the more I wonder whether such a study is really warranted, especially in Morocco. In my opinion, it would be far more interesting and rewarding to make . . . a study of this kind in Africa, south of the Sahara, where people eat a lot of fish. The only problem is that it would be necessary to restore the fish taste to the FPC which we have now managed to make beautifully bland . . . You may want to sound out AID in Washington about their feeling in this matter.[63]

Chapter 8
FPC in Chile: Contrasting Development Initiatives

Governmental policies concerning health and nutrition have been in existence in Chile since the 1920s. The country's first social security legislation, passed in 1924, established a national milk distribution program which has continued, with many modifications, to the present day.[1] In 1937, legislation that further expanded the milk program also gave special emphasis to maternal and child health care, with attention focused on preventive rather than curative medicine. Chilean government concern for the nutrition of vulnerable groups (chiefly manifested through food distribution in the milk and school feeding programs) was, therefore, not a new occurrence when the first FPC initiatives took place in the early 1950s. Protein deficiency in infants and children was a well-documented and publicized fact in the Chilean health sector during this period, and malnutrition was considered possible as a protein deficiency disease by nutrition experts in the scientific community (and consequently by the international agencies concerned with food and nutrition).

In 1953, interest in supplementing the protein intake of nutritionally vulnerable groups led the Food and Agriculture Organization (FAO), in collaboration with the National Health Service (NHS) of Chile, to conduct acceptability tests in Santiago of products made from South African fish flour. FAO had experience in the successful use of fish meal products in animal nutrition, and "in the search for cheap protein–supplementing products [for humans] . . . focused its attention on fish meals, which in a refined 'safe' form must be considered to be at least as useful for human beings as commercial fish meal has proven to be for farm animals."[2]

The "fish flour," supplied by the Marine Oil Refineries of Africa, Ltd., was an FPC which had been made from crude fish meal that was nine months old before the processing occurred. The meal had been solvent extracted and further treated to eliminate odor and waste. The FPC was used in various recipes in amounts varying from five to ten grams per serving. The FAO report of the tests simply states that "the dishes were eaten by 19 persons who in general found them acceptable. In some of the dishes a sandy texture was noticed." In another test "bread with 7 percent South African fish flour was tested in a group of 140 school children during fifty days with successful results. No objections were made to the taste of the bread which was found acceptable by all the children who took part in the experiment."[3]

The FAO tests were small in scale, not scientifically rigorous, and hardly conclusive. They simply proved that a very small group of people had not objected on the basis of taste to foods in which FPC had been incorporated. Yet this simple fact was received enthusiastically by nutrition advocates in Chile as it was in many countries where small-scale feeding tests with FPC were carried out. (See chapter 10.) The idea that this product was acceptable in terms of taste, coupled with the idea that the vast resources of the sea could provide a seemingly endless supply of raw material, created great interest in both domestic and international agencies. The apparent simplicity and accessibility of this "solution" to the protein shortage was its greatest appeal, but the effects of such oversimplification were not insignificant.

THE QUINTERO PROJECT (1956-1965): THE IMPACT OF FOREIGN ASSISTANCE ON NATIONAL NUTRITION EFFORTS

In Chile the motivations of the National Health Service (NHS) for entering into an FPC development project were closely related to the national milk distribution program. The National Health Service was created in 1952, an election year, by the amalgamation of the various government agencies responsible for health care in Chile. Just prior to the inauguration of General Carlos Ibañez del Campo, an authoritarian who promised a populist government, Chile's first nationally owned milk-processing plant was inaugurated. Administered by the NHS, the milk program, whose main beneficiaries were infants, preschool children (two to six years old), and pregnant and nursing mothers, underwent rapid expansion. "In 1951 630,000 kilograms of powdered milk were designated for some 32,000 beneficiaries, slightly more than 10 percent of all children in the eligible age category."[4] By 1954, the amount of milk distributed had doubled. Government imports of milk increased to meet the needs of the program, but opposition to the program then grew among domestic milk producers who feared reduction of their own sales. While the milk producers were not politically significant themselves, they were backed by powerful rural landowners represented by the National Agrarian Society, whose impact on Chilean politics was significant. The National Health Service in turn looked to FPC in the hope that the product would prove to be an alternative, and less costly, source of protein than milk. The fact that production of FPC would introduce new technologies and thus diversify Chile's underdeveloped fisheries was not important in the NHS's considerations. The fishing industry's main contribution to the economy was as an earner of foreign exchange through the export of fish meal and oil, and industrial growth was encouraged only in these areas.

UNICEF's decision to pursue FPC development in Chile evolved from its previous work in milk distribution, which had been a prominent feature of its postwar relief activities in Europe. As milk surpluses declined, UNICEF became involved with equipping dairy plants and by the mid-fifties had expanded its interests in nutrition to include the development of high-quality protein foods for children in locales where milk was not readily available.

The concerns of the NHS and UNICEF thus merged and both parties signed an agreement on February 16, 1956, "(a) to establish a plant for the production of *edible fish flour* at Quintero in the province of Valparaiso, Chile; (b) to provide a *nutritive high protein product* for *free distribution* to needy mothers and children under the supplementary feeding programmes now being organized by the Government."[5] Of these two objectives, UNICEF was more concerned with the former and the NHS with the latter. Unfortunately neither agency was concerned with the most important aspect of the project: the linkages—technical, managerial, and economic—between the two objectives. Both UNICEF and the NHS approached the production of FPC, in fact, as if FPC were a food, rather than a food ingredient. This critical misconception suggested that once the food was produced, it would be ready for government distribution to target groups. The real cost, in time and financing, of moving from a protein ingredient to a food product in which the protein was incorporated, and from plant to distribution and consumption was neither considered nor understood.

The stated objective of constructing a plant capable of producing a nutritive product to be made available for free distribution implies that the process chosen for use in the plant was capable of producing a nutritive product and that there was reason to believe that production would be economical enough to make free distribution feasible for the government. Though UNICEF acknowledged that the plant was to be a prototype for other plants, and in this sense was a test, the documentation related to the project implies a confidence in the production process that was unwarranted by experience.

The process selected for extracting FPC in Chile had been developed by a Swiss engineer, V. Dabsch, who had approached FAO with his idea in 1953 and was referred to UNICEF.[6] The only documented trials of the process prior to construction of the Chilean plant took place in Berlin in the pilot plants of two engineering firms with whom Dabsch was affiliated.[7] The first firm, Rud. Hartmann A. G., withdrew after initial tests and was replaced by the Borsig Company, which fabricated a new pilot plant for the Dabsch process.

The acceptable protein efficiency ratio (PER) of the FPC produced in the Borsig pilot plant[8] seems to have been enough to lead UNICEF to place a contract with the Borsig Company for the design and fabrication of a plant at Quintero, Chile, to produce FPC according to the Dabsch process. There is no information available that indicates that any other process was considered in comparison with the Dabsch process. Even if no other process was available to UNICEF at the time, it seems that prudence would have warranted much further work on the Dabsch process at the pilot level before proceeding to construction of a one ton per day (output) plant in Chile. This kind of work was not done, in fact, until after the first runs of the Quintero plant had failed. It must also be noted that prior to this time Borsig had not done any engineering work involved with food or food ingredient production; its principal area of design had been locomotives and locomotive parts.

In addition to Borsig, UNICEF, and the NHS, there was a fourth actor in the Quintero project, the Industria Pesquera de Altamar (ISESA). This Swiss fish-processing firm already had a factory at Quintero that was used for shrimp processing (up to and including canning) and for fish meal production. The choice of ISESA and of the Quintero site was made by UNICEF officials.[9] No evidence is available that indicates any link between the Swiss firm and the Swiss process engineer. Some evidence is available, however, that Chilean officials did not approve of the choice of site, partly due to doubts about the supply of raw material and harbor facilities at Quintero, and partly because ISESA was not a Chilean firm.[10] Whether these doubts were not expressed vigorously enough or were simply ignored by UNICEF is not clear. In any case, the choice of ISESA and Quintero held.

According to the Government of Chile–ISESA agreement, the FPC plant would be operated by ISESA's technical staff on the premises of the existing factory. The firm would charge the government the cost price for the FPC produced, with the reminder that "this contract is concluded for non-profit purposes and as a contribution by ISESA to the government's campaign to ease and solve the problem of providing food for children and needy persons. Though the plant would not become the property of the government of Chile until proved effective (that is, until an acceptable product was obtained), the NHS would be responsible for the administration of the project and would report regularly to UNICEF. UNICEF would provide the plant, the process, and the engineer to oversee the operation. The National Health Service would supply all other financial and material inputs.[11]

Construction of the Chilean FPC plant was completed by 1958, two years after the contract had been signed. Because the process engineer's arrival in Chile was unaccountably delayed, the major part of the erection of the equipment was done by ISESA staff without supervision. When the engineer did arrive in Chile, extensive, costly, and time-consuming revisions had to be made.

Three trial runs of the plant were attempted in 1958, and none was successful. The major difficulties were mechanical, but the consultant sent by UNICEF to observe the plant expressed more significant doubt about the process itself than about the extensive mechanical failures.[12]

The Quintero process was distinguished by a dual-phase extraction. The first stage was conducted with heptane or hexane and was designed to defat and dehydrate the raw material. The second stage employed ethyl alcohol and was intended to deodorize the product. The process engineer contended that ethyl alcohol was not used throughout the process for reasons of economy. Experience in the plant proved, however, that hexane was a highly flammable solvent, that it was more expensive than the alcohol locally available in Chile, and that it introduced difficult problems of toxicity in the final product. Though difficulties with hexane extraction were to continue throughout the life of the plant, no decision was ever made by UNICEF to change the process employed.

Such a decision would have required that UNICEF conduct the basic testing and evaluation that should have been done before the plant was constructed. This kind of technical expertise was clearly beyond UNICEF's capabilities and thus, once the process choice had been made and the plant installed, subsequent decisions concerning the process and equipment were simply attempts to cope with an inadequate technology.

It must be noted here, however, that after the 1958 trials a small pilot plant modeled after the Quintero plant and purchased from Borsig by UNICEF was set up in Woburn, Massachusetts. A program of process testing and development was planned under a grant made by UNICEF to the Massachusetts Institute of Technology. The work done at the MIT plant was focused mainly on dehydration, choice of solvent, and conditions of extraction. It was found that a simultaneous dehydration and extraction process with one solvent was far more effective than the dual-phase, two-solvent process used in Quintero. The various findings of this small testing unit, however, had little or no effect on the subsequent course of events in Quintero. The basically critical opinion of the MIT evaluators (members of the Department of Nutrition and Food Science) was not acceptable to UNICEF and the grant, which had been projected originally as a long-term one, was terminated before 1960.

The representative from the Chilean National Health Service, who viewed the first test runs in 1958 with the UNICEF consultant and noted the inadequate equipment and lack of a routine control laboratory, also pointed out what would prove to be of much greater significance in the future of the Quintero plant. He concluded his report by saying that "if the National

Health Service as well as UNICEF wish to obtain sound information, scientific and practical, it is necessary to contemplate the appointment of a trained professional with full powers regarding the industry men who will perform the exploitation of this 'Future Industry.' "[13] He was aware at this early stage that the allotment of central administrative responsibility for the Quintero project had been virtually ignored. The absence of a central force or planning body persisted throughout the history of Quintero, making it impossible to bridge the gaps between technicians at the plant, bureaucrats at the NHS, and policymakers at UNICEF.

By mid-1961, a full three years after the first trials, the equipment had been redesigned sufficiently to attempt new tests. Several of these were completed and some FPC was produced, though at the beginning of these tests no ethyl alcohol was available for use in the extraction process, other than that left over from the trials three years earlier. The NHS did not have sufficient funds to purchase a new supply of alcohol and finally the UNICEF office in Chile advanced the funds.

About 2000 pounds of FPC were produced, samples of which (totaling 60 kilograms) were sent to both the National Health Service and UNICEF-Santiago. Most of the FPC, however, was left in the Quintero plant. Although, according to the UNICEF report, "an analytical service was established at the University of Chile through the SNS [National Health Service] to aid during the tests . . . the work took too long to be of any practical help."[14] After these trials it was decided by the National Health Service that "raw material, plant control, and sampling will be carried out by an SNS inspector to be stationed permanently in Quintero."[15] Unfortunately, this decision, which would have been a step in the right direction for the Quintero plant, was never carried out because no direct action had been taken by UNICEF or the NHS since the 1958 trials in terms of allocating central administrative control of the plant.

Further trial runs were undertaken at the end of 1961 and beginning of 1962 when difficulties were again encountered with hexane extraction. While these tests were overseen by a Borsig engineer, there is no evidence that any NHS official was present. This absence was not due to poor planning but to the fact that National Health Service had by this time lost interest in the FPC venture and was seeking to rid itself of the official burden of the plant by transferring its administration to another Chilean agency.

The NHS had apparently realized that the technical aspects of FPC production would require considerably more time and money than had been anticipated. The investment of the Service had increased from the originally projected 20,000 escudos to 70,000 escudos (1 escudo = US $1).[16] A strong bias against ISESA had also developed within the NHS due to alleged inflation of prices. In its invoices to the NHS for the test runs of 1961, ISESA had not taken into account the fact that excessive consumption of water, steam, solvents, and electricity had been caused by an operator's oversight and by poorly insulated pipes. Whether the inflation was real or not, the NHS was convinced that the Swiss firm was not maintaining the nonprofit spirit of the original agreement.

ISESA's lack of interest in the efficient running of the plant is simply evidence that this small commercial firm, which had thought that it could profit in the not-too-distant future from involvement in a government food distribution program, had realized that it was dealing with a new technology rather than an attainable product. ISESA's business was fish meal, not product research and development, and it did not have the interest, the personnel, or the funds to pursue this project.[17]

The minutes of the meeting of the National Health Council (the executive branch of the NHS) of December 13, 1961, provide a historical summary of the involvement of the NHS in the Quintero plant and conclude that: "... the Service, in accordance with the Plan for Agricultural development of the government, believes that the country would be able to produce enough milk for its programmes, supply the children and mothers with proteins of a lower cost than that of the fish flour under the Agreement which had been estimated by ISESA."[18]

It was obvious that a protein shortage still existed in Chile, but the five-year history of technical difficulties and escalating costs at Quintero had turned the FPC plant from an exciting possibility into an unwanted burden. The National Health Service was more inclined to concentrate its efforts at this time on a known product and project (the milk program) than on what had proven to be, in the five years of its duration, an effort fraught with unexpected difficulties. By this time, too, a new administration under President Jorge Alessandri, a political conservative, was in office. Although the national milk distribution program greatly expanded during the first year of his term, this was only by virtue of congressional legislation enacted prior to his inauguration. During the rest of his administration no further expansions in the milk program nor other significant government investments in nutrition-related programs were made.[19]

The decision of the National Health Service is surprising only to the extent that it did not come sooner. One of the most serious errors in the Quintero project was the fact that the Chilean agency which undertook responsibility for the FPC plant had no experience in the development of a new technology. NHS had neither the technical expertise nor the political clout to question the wisdom of the donating agency's plans; it accepted them willingly, as a matter of course. As events showed, that initial enthusiasm was misguided but the momentum of bureaucratic and contractual agreements proved more forceful than critical findings (for example, the MIT findings and the reports of individual UNICEF and NHS consultants) concerning either the process or the lack of administrative cohesiveness in the project.

Though the conclusion of the December 1961 meeting was that the NHS would "transfer the Administration of the Plant in question to another Governmental organization,"[20] this in fact did not occur for four years, principally because no other Chilean agency was willing to administer the plant. In August of 1962, the title to the Quintero plant for the production of fish flour for human consumption was transferred from UNICEF to the Government of Chile, represented by the NHS. This transfer was in accord with the letter if not the spirit of the original contract which specified that the transfer could take place when an acceptable product has been obtained.

In 1964, the head of the NHS requested the assistance of a fishery technologist from the Pan American Health Organization (PAHO) principally for the purpose of establishing a laboratory to test and approve the FPC which had already been produced at Quintero. (Some production had occurred in 1963, although exact amounts are not known.) After two months in Quintero the consultant submitted a lengthy report in which he concluded that "further operation of the plant is not economically justifiable, and it is not recommended that funds be spent in rebuilding or modifying the plant in an attempt to increase its efficiency."[21] He suggested that the plant could be operated in order to produce limited quantities for experimental purposes "if cost is not an important factor."[22] UNICEF officials complained about the consultant's report, ostensibly on the grounds that he had not done what he had been hired to do, namely, establish testing procedures. It is clear, however, that UNICEF, though no longer officially associated with Quintero, was embarrassed by the report's focus on the ineffective allotment of administrative responsibility and the critical lack of technical expertise and managerial skill which had surrounded the project since its beginning.

In 1965, Chile's State Development Corporation (CORFO), which had taken over the administration of the plant from the NHS, arrived at a conclusion similar to that of the PAHO consultant and the Quintero project was terminated. The plant equipment, much of which was unusable due to corrosion and misuse, was disposed of by the Fisheries Department of CORFO.

In a very limited sense the FPC project at Quintero can be termed a success. In terms of the first objective of the UNICEF-NHS agreement—"to establish a plant for the production of edible fish flour at Quintero in the province of Valparaiso, Chile"—some FPC of acceptable nutritional quality was produced there. A rough estimate indicates that about 15 tons were produced from 1958 to 1964.[23] Some of this was used as enriching material for bread in an FAO-UNICEF applied nutrition program in the Salamanca area in 1962.[24] Some was sent to Peru for use in feeding studies at the British-American Hospital in Lima (see chapter 10).[25] Some of the material produced in 1964 was used in NHS feeding trials in Illapel.[26] Finally, in 1968 some of the remaining Quintero FPC was used by the NHS in the production of a baby food made from powdered milk, toasted wheat germ, and sunflower seed. This product, called Leche-Alim, was tested as a possible substitute for the milk distributed by NHS. It was hoped that Leche-Alim, being strictly designated as a baby food, would reach its target population more effectively than milk, which was often distributed throughout the whole family. The Leche-Alim experiment never got beyond that stage, however, apparently because mothers in the test showed a marked preference for milk as a baby food.[27]

In terms of the second objective—"to provide a nutritive high protein product for free distribution to needy mothers and children under the supplementary feeding programmes now being organized by the Government"—the project was a distinct failure. Though both UNICEF and the NHS had had considerable experience in food distribution programs and UNICEF in the funding and establishment of dairy plants, neither agency had ever had as its mandate the development of a new technology to produce a new food ingredient and the incorporation of this ingredient in acceptable vehicle foods ready for distribution. UNICEF's plan for a prototype FPC plant in Chile thus represented a fundamental misunderstanding of the conceptual and chronological gap between the idea of a technically feasible protein supplement and its actual implementation as a nutrition intervention.

The assumption implicit in the way in which the Quintero project was approached, both by UNICEF and the NHS, was that production of the product would be similar to production of an already known food product like dried skim milk. UNICEF looked on this endeavor with the hope that it would provide a new vehicle for food aid programs. The NHS hoped that with Quintero FPC it could quickly achieve a solution to the problem of the high cost of its milk distribution program. Because the technology was available, and the product was considered desirable, questions concerning the relevance of this kind of technology and its development requirements to the target populations in question—needy mothers and children—were never asked.

THE IFOP EFFORT (1964–1973): DOMESTIC DEVELOPMENT PROGRAM IN A CHANGING POLITICAL CLIMATE

Another international agency was involved in the origins of the second FPC endeavor in Chile, although the actual research and development efforts were a Chilean agency program. In 1963, FAO funded a $5 million grant to implement the formation of Chile's Institute for Fisheries Development. Prior to this an FAO mission had begun the first systematic and critical assessment of Chile's fishery development prospects. The original purpose of IFOP was to improve the use of existing fisheries in Chile by creating a strong technical base for rapid development. No thought was given at that time to developing new technologies for alternative uses of fish as food.

During the course of the FAO grant (1963–1968), a strong bias, amounting almost to a prohibition, existed against research on FPC for human consumption. The bias was based principally on FAO's estimation of the Quintero experience. Several members of IFOP's Department of Technology were food technologists, however, and interest in the area of fish food technology, and specifically in an FPC process, grew as IFOP developed. During the period of the FAO grant some research was directed toward the development of a fish-based milk substitute for weaning calves, with undisclosed interest in an FPC for human consumption that would use the same raw material and early process stages as the calf food.[28] In 1968, the FAO grant period ended and IFOP became a semiautonomous government agency under the administration of CORFO and the Ministry of Agriculture.

After the withdrawal of FAO, work at IFOP was openly directed toward development of an FPC for human consumption. The process of FPC production that was developed at IFOP was different from that of Quintero and that of the U.S. BCF. The latter two processes used chemical extraction with one or two solvents, while the IFOP process was biological, involving enzymatic hydrolysis. The great advantage of this process was in its economics since it required a fraction of the equipment involved in the chemical extraction processes. The product resulting from the enzymatic process was a bland white powder, nonhygroscopic, practically odorless, with no fish taste. It was soluble in water and, once dissolved, stable. Because of its solubility the IFOP FPC could be incorporated into many more foods than the BCF FPC and was thus significantly more useful than the U.S. product

By 1970, IFOP had been guaranteed 2 percent of the national copper revenues (about $1.5 million annually) and began to amplify its research substantially by funding FPC-related research projects in various Chilean universities and laboratories. Until 1969, IFOP's work on developing FPC for human consumption had been seen as an end in itself, namely, the development of a new technology that would utilize the nation's fishery resources. By 1970, however, a nutrition focus began to develop due in great part to the political climate of the newly elected administration of Salvador Allende, socialist candidate of the Popular Unity coalition. Allende, a doctor and a former Minister of Health, was an advocate of improved and expanded government health care. FPC researchers, who had the support of prominent nutrition advocates in Chile, sought increased financing from CORFO, the state development corporation which was now responsible for developing new industries and managing the socialized sector of the economy, in order to create a special center for FPC research. The main arguments made to CORFO were that FPC could be used to increase the protein intake of children in Chile and that it had export possibilities.[29]

Though the researchers had developed a usable protein supplement, which was exhibited and tasted in crackers and soft drinks, serious thought had not yet been given to what products FPC should be put in, and why. At this point a decision was made to focus FPC work on the development of milk substitutes, which could relieve the severe financial pressures the national milk program was experiencing at the time.

Salvador Allende's most popular campaign promise in 1970 had been "a half liter of milk per day" to every Chilean child under fifteen years of age, a promise the newly elected Allende government was trying to keep, though at a rapidly increasing cost. With the election of a socialist, USAID had withdrawn its donations to the milk program, which had been substantial. The National Health Service imported over 30,000 tons of dried milk in order to implement the program in 1971 and there was every indication that this amount would have to be increased. From 1971 to 1972, the price Chile paid for milk on the world market rose from $450 to $860 per ton.[30] Promoters of FPC work at IFOP suggested that an alternative solution to the needs of the milk program would be milk analogues based on FPC. While this idea interested CORFO, the National Health Service still considered that milk and the milk program were the best nutrition vehicles around.

In early 1972, the Industrial Protein Group (CPI) was formed within IFOP for the purpose of moving FPC-based products from laboratory to industrial scale. Various pilot projects were undertaken involving the construction of pilot plants and the utilization of existing industries.

By mid-1972, it became possible to produce and test the performance of a milk substitute for calves called APHL (based on FPC) which could suitably replace a product called Denkavit which was being imported at the time. In addition, CPI performed pilot operations that showed the technical feasibility of manufacturing, on an industrial scale, a hake pulp which could replace meat in sausages and an FPC-based milk substitute for human consumption.[31]

The National Health Service, having observed the variety of applicable products that CPI was developing, became very interested in the group's work and urged CORFO to invest in it. At the time, CORFO also was considering a proposal from Astra, the Swedish corporation, to produce a dehyddrated press cake from fish in Chile, which would be shipped to Sweden for processing into FPC, and then shipped back to Chile. CORFO considered having Astra work on anchovy in northern Chile, while CPI worked with the hake resource in the South.[32] Conflict within IFOP over which project to support was not based on technical preferences but had to do with political allegiances among the technologists and agencies involved. The Astra proposal had the support of those with communist allegiance while the CPI group had the socialist support. Because a suitable agreement could not be reached, Astra eventually withdrew its proposal and CORFO agreed to support the CPI venture.

CPI proposed that an industrial protein complex be established to produce a variety of products from a variety of raw materials. The actual mix of products would depend at any given time on local commodity availability and prices on the international market. The major product would be a powdered milk analogue based on a hake FPC and estimated to be approximately half the cost of milk. The hake offals would be used to make a milk replacement for calves, with the purpose of increasing the total milk supply for humans. At the time, 180,000 calves were being slaughtered each year because Chile could not afford to divert milk from human to animal use.[33]

The complex would also produce a hake pulp to be used as a meat substitute in sausages. It was estimated at this time that Chile was importing $32 million worth of meat yearly for sausage production.[34] Other intermediate products were also projected for the protein complex. It was estimated that the complex would produce 7500 tons per year of dried milk analogues for human consumption and 7500 tons per year of calf milk replacer. These figures would vary with the production of alternate products such as fish sausages. The input required for this production was 20,000 tons of hake. The estimated annual foreign exchange savings from the proposed plant were on the order of $5 to $7.5 million.[35]

In 1973, while sources of funding were being sought, Gonzalo Donoso, a member of Chile's national planning commission who was also a nutrition scientist, urged the CPI group to work simultaneously on a milk substitute based on milk and wheat and excluding fish. Donoso felt that the protein orientation of the milk program and of CPI's work was misguided in its lack of concern for the provision of calories.[36] He suggested that a milk-wheat product would be cheaper than a fish-based one and could be put into production sooner by using existing milk plants during slack periods. In response to this advice a product called FTL was developed by CPI from wheat protein concentrate, dried milk, and vegetable oil. It was designed for adults and children, to be used as a dried milk extender of breakfast and snack foods, both hot and cold. It was estimated that production of 20,000 tons of FTL (doubling existing dried milk production) could save $14 million a year in imports.[37]

Although the protein complex was close to becoming a reality, organizational and political changes occurred in 1973 that destroyed the bureaucratic stability the project needed in order to move to the construction stage. In the early part of the year CPI was made part of CORFO's Institute of Technological Research (INTEC) and its administrative link with IFOP was terminated. The reason for this shift is not clear, though it may have occurred in order to move CPI's administration into an institute ostensibly oriented toward research, rather than leaving it in IFOP, whose official business was fisheries development.

But INTEC was not, in fact, as interested in original research as it was in government service. Its definitely socialist political affiliation caused some problems for the CPI whose physical headquarters still remained on the premises of IFOP, a communist controlled organization. The CPI group was basically apolitical and this lack of affiliation to one group or another left it practically powerless. But though the CPI group lacked authority, the idea of the protein industry itself had taken hold.

By early 1973, technical and financial feasibility studies had been completed and CORFO had committed itself to the local currency costs of financing the project. CORFO then began to seek hard currency financing from several international sources including the World Bank, the Ford Foundation, and the Swedish government. The capital required for starting the new industry was estimated at $5 million, of which $2.8 million was sought in hard currency loan.[38] Technical and management consultants from these organizations were scheduled to visit Chile in October 1973 to determine whether support could be provided.

The sudden and radical change of government in the September 1973 coup caused the removal from office of most of the personnel directly responsible for the protein complex. CORFO was to make no further investments in state-run industries. The national milk distribution program was maintained but on a vastly reduced scale. The plan for an industrial protein complex based on the development of a variety of FPC products was put aside and has not been taken up again.

Vestiges of the IFOP–CPI endeavor remain, though, ironically, they do not involve fish. FTL, the wheat-milk product developed as a response to the demand for a cheaper, simpler, caloric milk extender, is now available in Chile under the commercial name of Superchil.[39] Another product, Fortesan, was developed at the same time as Superchil, but at the University of Chile. Made from bulgar, soy flour (fat free), soy oil, and powdered milk flavored with cocoa, this product was designed as a milk replacement for infants and as a protein supplement for older children.[40] Fortesan and Superchil are now being used by the NHS in government feeding programs.

The most important aspect of the CPI endeavor, which allowed the project to develop as far as it did, was an active government commitment to the improvement of the nutritional status of the population. Because this commitment existed beyond the level of rhetoric, an unusual and productive coalition occurred among technologists (IFOP–CPI), nutrition experts and advocates (NHS), and financiers (CORFO). The protein complex project was certainly not free from conflicting bureaucratic goals, but the commitment to improved nutrition, at least for a short time, put these conflicts in a secondary position.

Another aspect of the CPI effort that further differentiates it from other FPC projects is the basic concern, not with FPC technology per se, but with the development of acceptable and nutritious food products that would incorporate FPC and be economically feasible in terms of production. The means of marketing the projected products of the multiple protein complex were less clearly set out than the means of production, however. It was envisioned that the milk substitutes for human consumption would be used in government feeding programs in products that would, presumably, still be free to the consumer. But the effects of pricing policies for the calf milk replacers and meat substitutes, which would be sold to industries and reach the consumer indirectly, were not estimated prior to production. Since the complex did not become a reality, we cannot know how this government production venture would have finally affected nutritional status.

THE USAID PROJECT (1966-1969): A SEARCH FOR A PRODUCT OUTLET

Between the Quintero project and the CPI multiple protein complex another phase of attempted FPC development occurred in Chile, the USAID program. At the end of 1966, AID had been designated by the U.S. Marine Sciences Council as the "lead agency" for funding and administration of a program entitled "Food from the Sea for Undernourished People of the World." The imposition of this role and program on AID from the outside (that is, by the executive branch) is reflected in the confused and haphazard manner in which the agency pursued the program in Chile.

As a first stage of the AID program a survey team was sent to Latin America in September 1967 to evaluate countries as possible locations for FPC activity. The objectives of the survey were, in order of their importance,

to determine which of the countries visited would provide the best opportunity for the introduction of a commercial Fish Protein Concentrate system;

to determine whether any of the countries visited had an adequate research program that could utilize a small pilot plant capable of producing FPC . . . by an alcohol extraction process;

to observe present patterns of fishing activity and the marketing and consumption of fish in the countries visited; and

to make other observations relevant to FPC in Latin America.[41]

The overall recommendation of the survey team was that Chile be chosen for the FPC demonstration project. The team also concluded, however, that the provision of an FPC pilot plant to a country in Latin America should be deferred since no country visited had a comprehensive FPC research program already developed.[42] While Chile had the engineering capability for the construction and operation of an FPC plant, the report emphasized that the design of this plant should be based on experience gained from the U.S. demonstration plant which, at that point, had not yet been constructed.[43]

It is thus apparent that the immediate goal of the Food from the Sea Program, whatever the objectives of the survey team or the feasibility study, was to find a country with adequate experience and infrastructure for the distribution of the FPC that was being produced under contract with AID by Alpine Marine Protein Industries. The agency could thereby fulfill part of the mission that had been given to the Food from the Sea Program, namely, distribution of FPC to developing countries. Had AID provided a pilot plant to Chile, there would have been little incentive for that country to import the Alpine product.

As a result of the initial survey, Chile was selected by AID to be the subject of a study concerning the feasibility of commercial FPC production in Chile. The Inter-Ministerial Protein Advisory Committee (CIAP), which had been formed in 1966 to consider possible solutions to the country's protein deficit, negotiated with AID concerning the FPC program. A misunderstanding arose at the beginning of these negotiations in 1968 which created confusion and mistrust that was to leave its mark on future negotiations. An AID official conveyed the impression to members of CIAP that what was being offered was, indeed, a pilot plant for the production of FPC within Chile, prior to the construction of a full-sized plant. This was definitely contrary to the conclusion of the survey team that the provision of a pilot plant should be deferred. This misunderstanding persisted throughout the year, for the minutes of the CIAP meeting of December 20, 1968, state that the Committee "must insist that AID fulfill its offer of February of this year of a small pilot plant to produce in this country fish protein concentrate for human consumption, in order to have the product until the industrial plant is installed."[44]

By February of 1969, it had become clear that a pilot plant was *not* being offered but that 250 tons of FPC produced in the United States by Alpine Marine Protein Industries would be allotted to Chile, the first shipment of which was to be ready in June of that year. Of the 1000 tons originally promised to AID by contract with Alpine, only 172 were found acceptable by AID. Of these, about 100 tons were designated to be sent to Chile. At the end of July 1969, 500 kilograms arrived at the Santiago airport. Chilean bioassays of the product showed it to be of poor quality in terms of taste and mixing capability. The Chileans maintained that AID–Alpine product was inferior even to the FPC remaining from the Quintero plant.[45]

AID terminated its contract with Alpine at the end of 1969 due to the poor quality of the product. A memo of CIAP summarized its own situation with regard to AID (as of December 1969) stating that "all negotiations are paralyzed and the date of termination of the feasibility study concerning the installation of a plant in Chile for FPC for human consumption is unknown. Nor is the date of arrival of the 300 [additional] tons of FPC known, nor even if they will arrive."[46]

General Oceanology, Inc., a consulting firm in Cambridge, Massachusetts, had been awarded the AID contract to study the feasibility of commercial FPC production in Chile. Personnel from General Oceanology began work in Chile in the summer of 1968. Ignoring previous studies that had been done, the firm produced its own extensive examination of all factors relevant to the commercial production and marketing of FPC in Chile. Some Chileans in the Ministry of Health and CORFO objected to General Oceanology's procedures, arguing that the consulting firm was delaying unduly the establishment of the FPC industry project.

The final version of the General Oceanology study did not appear until June 26, 1970. Prior to this, however, events had occurred that would guarantee the end of the AID-FPC effort in Chile. Internal reorganization within AID, in line with the policies of the Nixon Administration, had led to the termination of the Office of the War on Hunger. The Food from the Sea Program, which had been part of this office, became part of the Technical Assistance Bureau (of the Office of Agriculture and Fisheries) and was significantly reduced in scope. At this point the AID-FPC program was in effect defunct since, with the cancellation of the Alpine contract, no supply of FPC was available and interest within the Agency in encouraging FPC development anywhere was minimal. Finally, with the election of Salvador Allende in the fall of 1970, AID stopped almost all forms of assistance to Chile, including its "FPC program" and, more important, its very real donations to the National Health Service milk program.

The General Oceanology feasibility study, though never implemented in any sense, reached interesting conclusions. The final report stated that commercial FPC production was possible "with major Chilean government participation in financing, in promotion and in extending possible subsidies in support of market development, domestically and internationally."[47] This kind of major government support was in fact the kind of support that the multiple protein complex of IFOP-CPI represented, though whether this support would have been sufficient to generate an industry was never proved.

The General Oceanology study had further concluded that "in certain circumstances the initiation of an FPC plant in Chile may be wholly justified in terms of a long range business whose primary objective is centered on interesting profit potentials over the long term rather than the short term considerations."[48] It is interesting that the only significant commercial production of FPC that exists today began precisely as this kind of "long range business, centered on interesting profit potentials over the long term rather than short term considerations." This is the work of the Astra Group of Sweden. The important difference between Sweden and Chile in this case is that the Swedish company is private and operates in a developed and protein rich country. It does not produce FPC on government subsidy for the improvement of the nutritional status of the local population, but rather sells FPC to an international development agency for donation to needy countries.

The constraints that limited attempts to develop FPC in Chile were insignificant in Sweden where no pressure existed to develop a product quickly in order to fulfill an urgent domestic need. It was the urgency of this need in Chile that blinded the National Health Service, during both the Quintero and

AID efforts, to the fact that a product like FPC would require extensive development before cost and acceptability constraints could be overcome and before products incorporating it could reach nutritionally vulnerable groups.

Both these efforts, though they nominally addressed the problem of protein shortage, in reality grew out of the needs of the initiating agencies rather than those of any malnourished population. In the case of UNICEF, the initiation of the Quintero project was the hopeful beginning of a new field of endeavor (subsequent to skim milk distribution) which proved to be far more costly and complicated than anticipated. In the case of AID, the FPC program in Chile was the foot-dragging fulfillment of an unwanted mandate. In both instances, the National Health Service hoped it could quickly obtain a product to further its own goals of food distribution. As the technological and economic constraints of development became apparent, however, the idea that FPC could easily solve the protein problem disappeared, as did bureaucratic enthusiasm for it. It was only in the case of the CPI multiple protein complex, where development involved a concerted effort to produce domestically appropriate and usable food products, that FPC came close, even in theory, to reaching nutritionally vulnerable groups in Chile.

Chapter 9
FPC in Sweden: Successful Development in the Private Sector

In each of the countries considered in the previous chapters the major part of FPC research and development was carried out either by government agencies or with government funding, and these endeavors, as we have seen, ended with attendant complications and the fairly consistent verdict of "failure." Sweden is the exception to this rule. It is the only country where FPC was developed exclusively within the private sector and where FPC is being produced today on a significant commercial scale.[1] Moreover, since this development and production have occurred under the aegis of only one company, the Astra Group, a small number of actors have been involved and little controversy has surrounded it. Astra's success and this lack of controversy stand in striking contrast to other commercial as well as governmental endeavors to produce FPC. In order to understand what Astra's success means for other food products similar to FPC and for nutrition policy formulation, we will take a brief look at the chronology of development and consider the nature of the corporation itself.

INITIAL INTEREST AND DEVELOPMENT OF PRINCIPAL PRODUCTS (1957-1970)

The Astra Group, established in 1913, is Sweden's largest pharmaceutical corporation and one of the top forty such companies internationally. Astra also manufactures chemical products for industrial, agricultural, institutional, household, and leisure use, and more than half of its sales are in international markets.

Astra's work with FPC began in 1957. The company's experience with hospital needs had created an interest in developing a high protein food for hospital use. It was thought that an inexpensive storable protein concentrate from fish could be produced for this purpose and by 1960 Astra researchers believed that an appropriate processing method had been attained on a laboratory scale. A pilot plant was then established at Bua on the West Coast of Sweden.[2] This was the informal beginning of Astra's nutrition division. Technical difficulties were encountered in transferring the product from laboratory to pilot plant scale,[3] but these were eventually overcome and in 1963 Astra's FPC for human consumption was approved by the Swedish Food Administration.

Work at the pilot plant continued until 1966 when a full-scale FPC production plant with an output capacity of approximately 12,000 tons per year was constructed at Bua. Astra's nutrition division had two commercial goals: to develop an animal-grade fish protein as a substitute for powdered skim milk in calf and other animal diets, and to develop a human-grade FPC for use in food, the principal target area being developing countries. About this time a new nutrition research laboratory was built at Mölndal which included extensive chemical analysis equipment and animal feeding facilities. By 1967, the Astra's nutrition division employed about four hundred people,[4] and Astra had two principal FPC products, "Prot-Animal,"[5] a substitute for milk protein used in calf milk replacers and specialized animal diets (for example, for laboratory animals), and "Astra Protein" for human use.

The FPC production process developed by Astra, like the U.S. process, uses solvent extraction with isopropyl alcohol. Astra FPC is made from whole fish, but until 1973 eviscerated fish had been used to comply with then-current U.S. FDA regulations which prohibited FPC made from whole fish, contending that such a product was adulterated and consequently filthy.[6] When the FDA approved the use of FPC made from whole fish in 1967, the only species allowed was hake. The basic raw material of the Astra FPC was (and is) North Sea herring, so Astra continued to eviscerate its product in the hope that a separate food standard could be obtained for this product from the FDA since under such a standard the product would be considered a food ingredient rather than an additive. Astra discontinued its attempt to achieve this standard in 1970 when the FDA approved the use of whole herring in the manufacture of FPC.

Three points are worthy of consideration here. First, Astra is a research-intensive company whose principal endeavors have been in the area of pharmaceuticals. Long periods of time required for product development, and accompanying uncertainty, are the rule, rather than the exception, in pharmaceutical development. The duration of development time for FPC, like the cost of its development, were unknowns that Astra was prepared to bear. This diversified and well-capitalized company could undertake research and development of a high-risk product like FPC without the pressures of dwindling funds and accountability to a higher agency experienced by developers in the countries we have already considered. Astra, in fact, used its animal-grade FPC as a buffer product, sales of which supported development and testing of Astra Protein. Second, Astra's efforts to develop a process that would comply with U.S. regulations is indicative not only of the company's recognition of the international effect of FDA decisions, but, more important, of Astra's realistic awareness of and interest in an eventual U.S. market for its product. Third, Astra's development of FPC has taken place within a consistently favorable environment of noninvolvement by the Swedish government, with the exception of government purchase of FPC. Because the government was not involved in funding development, it had no interest in specific decisions made by Astra in the course of its progress, and Astra was thus not burdened by the need to please various political interest groups. In addition, Astra was favored by conducting FPC development in a country where fish is a principal food and has been for centuries and where the idea of making a new food component from fish was found to be unobjectionable by the public or, to be more accurate, was hardly noticed at all.

In 1967, Astra began discussions with the U.S. government and several U.S. companies concerning collaborative work on FPC. Astra was interested in supplying the process to be used in the second FPC plant projected to be built in the United States after the Experiment and Demonstration Plant was completed.[7] This second plant, however, was never constructed. Astra approached the U.S. Bureau of Commercial Fisheries about collaborative work but the BCF was deeply involved in its own FPC development and had no interest in working with a foreign company. The opinion of the Marine Sciences Council at that time, however, was that Astra Nutrition was "serious, progressive, and competent" and that the United States should "discretely encourage an Astra–U.S. firm association."[8]

For its own part, Astra knew that while it had developed a suitable process for the production of FPC, it did not have the expertise in food processing and development that would be necessary to formulate acceptable and marketable foods made with FPC. The company continued to seek affiliation with a U.S. firm.

During the late 1960s, field tests of Astra's FPC for human consumption began. Tests were made in Uppsala, Sweden, and in Teheran, Iran. Schoolchildren in both places were given FPC-enriched buns over a five-month period, with a favorable response. The first significant use of Astra FPC for human consumption occurred in 1968 when 20 tons of Astra Protein was shipped to Biafra and distributed there by the Lutheran Aid Society and the Swedish Red Cross.[9]

EXPANSION AND PRODUCTION (1970–1974)

In 1970, Astra found the kind of alliance it had been seeking in terms of marketing its food-grade FPC. In February of that year it joined with the National Biscuit Company in New Jersey to form the Nabisco–Astra Nutrition Development Corporation with each company owning 50 percent of the new group. The alliance with the U.S. firm had the following objectives:

1. to develop food products enriched with a high-quality protein derived from fish;
2. to develop markets for these products among groups of people in countries most in need of more dietary protein; and
3. to research and develop other newer sources of protein for use in developing as well as developed countries.

Another indication of Astra's faith in FPC and of the booming market for Astra's Prot-Animal in Europe (where Astra was the principal supplier of milk substitutes for calf feed) was the purchase, late in 1969, of a $7 million, 25,000 ton floating factory ship, the M.S. *Astra.* The factory ship, a converted whaler, was serviced by 11 fishing trawlers and had an annual production capacity of 40,000 tons of FPC. Since the processing plant at Bua was not in a prime location in terms of raw material availability, the *Astra* would travel to where the fish were available in abundance, process them after they had been caught by the trawlers, and bring the intermediate product back to the plant at Bua. Originally it was intended that a full production plant would be installed on the ship. It became impractical, however, to run an extraction plant on board and it was resolved that production would go as far as a storable intermediate product which would then be brought to the plant at Bua for final processing into Prot-Animal or Astra Protein. The optimism with which the company viewed the outcome of this endeavor is reflected in the words of the Group's President at the time the M.S. *Astra* was acquired.

> We are convinced that the problem of protein deficiency throughout the world is capable of solution in the next few years, and that protein extracted from fish is the best source. Now that we have solved the technical problems, we have thought it urgent to apply this know-how to a production that is adapted to large-scale consumption in different parts of the world.[10]

With the acquisition of the M.S. *Astra,* sales did increase markedly, but principally in the area of animal-grade FPC and fish meal (which the factory ship also produced) rather than in Astra Protein for human consumption. About this time Astra sought to expand its Astra Protein production areas and sales market by negotiations with the government of Chile, which was highly interested in FPC. These negotiations were discontinued when Chile chose to restrict itself to the FPC work of its own agencies.[11] In 1973, sales of Astra's Protein Products Division, representing Astra Nutrition (food-grade protein), Astra Prot-Animal (animal feed protein), and Astra Overseas Fishing (fish meal, principally produced by the M.S. *Astra*) reached $14.7 million, a figure that had steadily increased since the acquisition of the ship in 1970, where division sales were estimated at $5.2 million.[12]

Every FPC case history we have examined has had its peculiarities, and the Astra story is no exception, though in this case of successful product development the one "peculiarity" was significant enough to seriously affect the whole venture. On Easter Sunday morning 1974, the *Astra* was riding in calm seas off the coast of West Africa.[13] It was a fair day and the ship, accompanied by only one of its fishing vessels, had all its engines stopped. A freighter appeared on the horizon, heading on a course directly in line with the *Astra.* The ship's skipper and crew assumed that the freighter would alter its course as it came closer. Incredibly, the expected did not happen and the freighter (of Liberian registry) rammed the M.S. *Astra* broadside. The ship began to sink and the crew was evacuated to other vessels that had come to the rescue. One man, however, was missing. A salvage tug set out from Las Palmas to aid the damaged factory ship but did not arrive in time. The M.S. *Astra* sank on April 17, in almost the same location where it had been rammed three days earlier. The missing crewman was never found.

SEARCH FOR A MARKET (1974–1976)

Whatever the cause of the accident, the results were serious. One man had lost his life and the company's greatest source for raw material supply to its factory had disappeared. The loss of the M.S. *Astra* meant that raw material for the Bua plant had to be obtained, as before, by contracting fishing vessels, none of which could process its catch on board and deliver an intermediate product, as the factory ship had done. Sales of Astra Overseas Fishing, which had been forecast at $20.9 million for 1974, fell to $3.4 million.[14] In the group's annual report for 1974, the Prot-Animal division is not even mentioned as a separate division, evidence of decreased sales resulting from the loss of the ship compounded by decreased demand for milk substitutes in Europe, where milk production had increased considerably.

Sales of Astra Protein, however, were reported to have doubled (to $2.5 million) in 1974, reflecting the continuing commitment of the Swedish International Development Authority (SIDA) to supply FPC-fortified flour to the UN World Food Program (WFP). This commitment had begun in 1973 when the Swedish government allocated one million dollars for FPC enrichment of the flour Sweden donates to the WFP.[15] The Program had tested Astra's FPC in 1972 in approximately twenty countries and found it generally well accepted.[16] The Program received 24,000 metric tons of Swedish wheat flour fortified with 4 percent FPC for the 1973–1974 period and an equal amount was committed by SIDA for 1975–1976.[17]

In 1975, Astra converted its plant at Bua in order to decrease Prot-Animal production and increase that of Astra Protein. The annual capacity of production of the human-grade product was increased to 9000 tons per year.[18] A Group publication of January 1976 indicates the market toward which this increased production was directed:

> Even though the need for good quality protein products in the developing countries is quite obvious, the local purchasing power is often missing, and we can hardly speak of an existing commercial market.
>
> The purchasing power for products such as FPC can, however, be found in the national nutrition programs, financed and regulated by the local governments and/or in the international and national aid and assistance programs.[19]

But, despite these hopes, the increased demand Astra expected never came in 1975, nor in the early part of 1976. SIDA continued to be Astra Protein's major buyer. Although the Astra Group's sales for 1975 were estimated at $290 million, Astra Nutrition's sales for 1975 decreased to $1.5 million. This drop further reflects the declining market for milk-protein replacers for animal feed, but more significantly it demonstrates the absence of any significant interest in FPC as a vehicle for nutrition improvement in developing countries. As a result of decreased demand for FPC, surplus stocks have accumulated at the Bua plant to such an extent that the Astra board has decided to stop production of FPC until stocks are reduced to a normal level. In the meantime, activities of the Astra Protein sector are being concentrated on marketing and sales.[20]

Since commercial production of FPC began at Astra, a small part of this production has been sold regularly to the Swedish baking industry which incorporates it into breads available in retail stores. FPC-fortified products are also used by hospitals and military installations in Sweden. It is possible that Astra will increase efforts in this area in its attempt to bolster FPC sales. In addition the Astra Group has recently investigated the possibility of establishing an FPC production plant in the United States. While there is no significant demand for the product at present, Astra is interested in the possible uses of FPC as a protein extender in new food products based on vegetable proteins. Though one of the least expensive animal proteins (in terms of price per pound of protein),[21] Astra's FPC still exceeds the price of vegetable protein considerably. The price of Astra Protein in the United States today is seventy-five cents per pound, though Astra representatives estimate that this price could be reduced to between fifty-five and sixty cents per pound if a U.S. plant were established.[22]

While the World Food Program has not canceled its current commitment to using FPC-fortified flour donated by Sweden, it has not increased its order nor made any firm plans for its future use. The supplies of the flour already received have been used exclusively in the United Arab Republic in a number of large projects concerned with land development and settlement where the

World Food Program is providing supplementary food supplies.[23] Without the SIDA purchase, demand for Astra's food-grade FPC would of course be greatly reduced. Nor is the future outlook hopeful. The World Food Program is becoming increasingly interested in the use of a cruder, and consequently less expensive, FPC. This is a Type B product according to the classification of the UN Protein Advisory Group (PAG); it is nonextracted and has a higher fat content than Astra's FPC, which is Type A. This product, manufactured by a Norwegian company, is currently being tested by the Food and Agriculture Organization (FAO) and the WFP in a number of countries in West Africa and the Far East. A large quantity of Type B FPC is also being programmed into Bangladesh, where it has been under extensive testing for about five years.[24]

If the World Food Program finds Type B FPC as acceptable as the more expensive Type A, it will clearly opt for the less costly product, as would other international agencies. Astra might still be in a position to be a prime producer, but the details of their Type A process, designed and developed to produce a highly bland product, would no longer be necessary in that case.

Though the role of FPC as a protein supplement in developed countries is not yet a closed issue, the developing world has clearly not proved to be a lively market for FPC. Nabisco-Astra has developed various products incorporating FPC and these have been tested in government-funded feeding programs in the Mideast, but as yet no significant investment has been made in them by a government of a developing country. The acceptability of these products in terms of taste and other characteristics is not in question. Rather, the issue is whether governments in developing countries want to invest in a nutrition intervention that would require the importation of either an expensive product or an expensive technology.

Thus, the aspects of the Astra FPC development endeavor that have made it a success—the financial resources of the company, its experience in long-term development, and its technological expertise, all of which Astra was willing to commit—demonstrate that different technological problems can be solved and underutilized resources developed, given adequate inputs of time, money, technical expertise, and commitment. All this effort, however, has not created a sizable or clearly defined market. While the Astra FPC model is indeed exemplary in terms of process development and production, it is entirely questionable in terms of appropriate nutrition interventions for low-income countries.

Chapter 10
Worldwide Development of Fish Protein Concentrate

In order to complete the picture of worldwide interest in FPC, we have assembled in this chapter pertinent information on documented efforts in forty countries to introduce, utilize, and/or manufacture FPC. Together with the descriptions of FPC-related developments in the five countries already discussed, the following sketches, representing twelve European, eleven Asian, eight American, eight African countries, and Australia, provide evidence for the contention that FPC has indeed been the subject of significant global interest. These country summaries will also be useful starting points for further, more specific studies.

ANGOLA

Sometime in 1959 or 1960, the Food and Agriculture Organization (FAO) obtained a quantity of FPC from Astra Nutrition (Sweden) that was used for acceptability trials in several countries, including Angola. At the August 1960 meeting of the UN Protein Advisory Group (PAG) in Washington, D.C., USA, the FAO reported on tests that were conducted in Luanda by Dr. J. A. Pereira-Nuñez. Dr. Pereira-Nuñez was quoted as follows:

> The deodorized flour was first tried out in an establishment run on Public Charity in this town, and even from the very first, we noted repugnance on the part of those who tried out meals in which the flour had been used. This fact made us try out several kitchen menus, in the hope that the flour would prove more acceptable, if, instead of dried vegetables, which we first used, we should try out dishes based on fish, which are very much used in our

Portuguese cooking—fish soup, fish cakes, fish rice, etc. These fresh efforts also proved negative on account of the "sui generis," taste which the flour, although odorless and tasteless when uncooked, takes on as soon as cooked.[1]

FAO reported that it had found this "quite surprising and contrary to experience elsewhere," especially in view of the fact that samples of FPC sent to them by Dr. Pereira-Nuñez showed no signs of flavor reversion.

No subsequent mention of this or other trials in Angola appears in the PAG record.

References

1. Nutrition Document R.8/Add. 10, WHO/FAO/UNICEF/PAG, August 1960 Meeting, Washington, D.C., pp. 5, 6.

ARGENTINA

Some interest in FPC has only recently been evidenced in the technical literature.[1,2]

References

1. A. M. Sotorres, N. A. Suldrup, and E. L. Itten de Entizne, "Development of new fish protein concentrates in Argentina. I. Basic chemical composition," *Arch. Fischereiwiss.*, 22:2 (1971), pp. 130–135.

2. A. M. Sotorres, and H. Hoermann, "Development of new fish protein concentrates in Argentina. II. Solubility and fractionation of proteins," *Arch. Fischereiniss,* 22:3 (1971), pp. 280–286.

AUSTRALIA

FPC-related activity has taken place intermittently in Australia for some twenty years, beginning with commercial experiments in solvent extraction of animal products. In 1959, a U. S. patent (3,076,708) was issued to J. C. Cavanaugh of Australia covering solvent extraction techniques for the dehydration and defatting of meat, fish, coconuts, seeds, grains, legumes, and other vegetable matter. (The language of the patent was subsequently considered sufficiently broad, by the U.S. Patent Office, to cover the Bureau of Commercial Fisheries (BCF) process and thus to warrant denial of the Interior Department's application for a separate patent on BCF's isopropanol process.)

Specific FPC activity seems to have lapsed until 1971, when Fish Protein Concentrate (Tasmania) Pty., Ltd., embarked on a commercial venture in

Tasmania to make a high-grade fish meal for animal rations. The company had in mind eventually manufacturing FPC for human consumption, and proceeded with its plans despite misgivings of the Tasmanian Regional Laboratory of the CSIRO.[1] The status of the venture was summarized recently by the Director of Agriculture of the Tasmania Department of Agriculture.

A high quality fish flour was produced commercially from jack mackeral (*Trachurus declevis*) at Triabunna in Tasmania during 1973/74. Production . . . for 12 months was approximately 1200 tons from 5400 tons of fish.

The company . . . operated 2 purse seine vessels to supply their factory which utilized a heptane extraction process (French design) to produce the fish flour and fish oil. The FPC was distributed within Australia to various animal feed buyers and the venture was funded from private sources . . . FPC production ceased in June 1974 (the end of the mackeral season) and the plant has since closed down with the boats being used for other purposes (tuna catching in the main). With the company under receivership the exact status of future FPC production is unclear but it appears likely that the original plant will be dispensed with. A likely alternative lies in conventional fish meal and canned fish production.

Reasons for the failure of the venture are manifold and unclear. A major problem appears to have been overspecialization on the sole target of FPC production, together with logistical problems with plant and machinery . . .[2]

References

1. June N. Olley, unpublished report, Tasmanian Regional Laboratory, CSIRO. September 9, 1971.

2. A. R. Mead, Letter to Dr. June N. Olley, September 25, 1975.

BANGLADESH

Experimentation with FPC in Bangladesh predates the formation of that country (1971) and the initiation of the College Park, Maryland, program. It is difficult, however, to establish the beginnings of the FPC story in Bangladesh more precisely. It is known, for instance, that in 1959 Pakistan was attempting to produce an acceptable FPC,[1] and it is more than likely that this work was taking place in the Dacca laboratory of the Pakistan Council for Scientific and Industrial Research (PCSIR), rather than in the Karachi PCSIR lab.

Certainly the Dacca laboratory was engaged in a search for a cheap and easy method for commercial production of FPC early in 1961.[2] Although reports began to appear by 1965, giving some details of the FPC research taking place in Dacca,[3,4,5] an attempt by an AID representative in 1970 to elicit similar (though more up-to-date and detailed) information met with no success.[6] Some tentative commercial interest emerged in 1967[7] but no private manufacture of FPC in East Pakistan or Bangladesh seems to have been attempted.

During October to December 1971, acceptability trials were conducted among Bangladeshi refugees in Calcutta, and subsequently 60 tons of FPC were distributed in Sylhet, Kurigiam, and Lalmonirhat early in 1972.[8] Other trials were conducted at the Cholera Research Institute and the Dacca Shishu Children's Hospital and were followed by distribution of 1650 tons of FPC supplied by Norway.[9] The 1973 report of FAO's nutrition adviser outlined plans for subsequent large-scale feeding trials, which would require the acquisition of 5000 tons of FPC in 1973/1974, and urged the establishment of a domestic FPC industry in Bangladesh.[10] No FPC industry has been established, but as of January 1976, 5000 tons of Type B FPC (which has been under testing by FAO in Bangladesh for five years) had been pledged to the World Food Program for distribution there.[11]

References

1. Nutrition Document R. 8/Add. 10, WHO/FAO UNICEF PAG, August 1960 Meeting, Washington, D.C.

2. Report of the FAO Indo-Pacific Fisheries Council meeting, January 6–23, 1961, Section A, Processing.

3. H. N. De, and A. Hannan, "Fish flour. Biomedical and nutritional studies. I. Yield and chemical composition of fish protein concentrate (FPC) and fish meal (FM) prepared from various fresh-water fish of East Pakistan by application of improved techniques," *Pakistan Journal of Scientific Research.* (Dacca, Pakistan) 11:3 (1965), pp. 85–94.

4. H. N. De, A. Islam, and Y. Mollah, "Fish flour—its biochemical and nutritional studies. Y. Supplementary effect of fish protein concentrate (FPC) on the cereal proteins evaluated by determination of PER values and measurement of liver protein nitrogen, non-protein nitrogen and fat contents," *Pakistan Journal of Scientific Research,* (Dacca, Pakistan), 3 (1966), pp. 31–42.

5. H. N. De, A. Islam, and Y. Mollah, "Biochemical and nutritional studies on edible fats and oils in relation to fish flour protein utilization. VII. Influence of feeding cowghee, groundnut oil, and mustard oil with fish flour on the gains in body weight, protein efficiency ratio, and caloric efficiency ratio values," *Pakistan Journal of Scientific Research,* (Dacca, Pakistan), 5:1 (1968), pp. 12–31.

6. Norman L. Brown, "Survey of potential for fish protein concentrate manufacture in Pakistan—1970," Report to U.S. AID, 1970, p. 9. PCSIR scientists retained a personal financial interest in any commercial development of their

laboratory process; thus, they were reluctant to disclose any technical information to visitors—or even colleagues. The only information available from the Dacca laboratory in 1970 pertained to the proximate analyses of some of the variety of FPCs that had been made (including FPC from oysters!) and results of nutritive value studies.

7. In a letter to Shamsul Alam Khan, U.S. Consulate-General, Dacca, July 6, 1967, A. G. Hazari, Director, Cellux Pakistan Ltd., requested information on FPC from the U.S. government, indicating that his interest in this material as a source of protein stemmed from a similar interest on the part of the Medical Research Centre, Segunbagicha, Dacca. This letter was sent to the U.S. BCF a few days later, but there is no record of a response.

8. Cato Aall, "Some long and short term aspects in dealing with nutrition problems in Bangladesh," available as Information Paper No. 27, "Nutrition problems in Bangladesh," United Nations Relief Operation Dacca (UNROD), 1973.

9. *Ibid.* Norwegian Church Aid supplied 60 tons of FPC for the early trials in 1972. As part of its relief contribution Norway supplied 650 tons of FPC in October 1972 and 1000 tons in March 1973.

10. See note 8.

11. Wolfgang Krone, Senior Fishery Industry Officer (Fish Utilization), Fish Production and Marketing Service, FAO, personal communication, 1976.

BRAZIL

Interest in FPC in Brazil has been confined to process development in government laboratories. For approximately ten years a small program has been underway at the Instituto de Pesquisas da Marinha (Marine Research Institute) in Rio de Janeiro, primarily as a result of the persistent efforts of its director, Admiral Paulo Moreira da Silva. With the cooperation of USAID, FPC development at the IPM moved from bench-scale experiments with an isopropanol extraction process to a small pilot plant. The U.S. agency provided engineering assistance and equipment from the College Park laboratory of the National Marine Fisheries Service (NMFS) (formerly the BCF).

Recently a move has been made to interest industry in the manufacture of FPC. The Companha Nacional de Alimentacão Escolar (National School Lunch Company) is embarking on a project to supplement the lunch of each schoolchild in the program with 10 grams of FPC, starting in March 1976. The CNAE has ordered 10 tons of FPC from the IPM—presumably to be made in its new pilot plant facilities—and is using this program to stimulate commercial interest.[1]

References

1. Private communication to Dr. N. L. Brown.

SOCIALIST REPUBLIC OF THE UNION OF BURMA[1]

Burmese experience with protein concentrates made from fish dates from at least twenty years ago. Because of the liking for foods and condiments with a strong fishy flavor traditional in the Burmese diet, however, the early experiments with "fish flour" were actually experiments with a carefully prepared, sanitary fish meal.

While conducting nutrition surveys from 1954 to 1957, Wil Pranger and Mary Ross, FAO nutritionists, also prepared a quantity of flour from dried sandfish.[2] This fish flour was then used in acceptability tests with 140 two-and-a-half to six-year old children in a nursery (for twelve months), 30 pregnant and lactating women (for nine months), 50 men and women in a home for the aged poor (for one week), and 800 hospital patients (for one week). The fish flour was added to many soups, sauces, green salads, and other vegetables, commonly eaten by the Burmese with *ngapi,* a fermented fish sauce. The nutritionists concluded, "There remains little doubt of the acceptability of fish flour, used in 20 gramme amounts, in Burmese dishes."[3] In a subsequent report commenting on the work of Pranger and Ross, Dr. H. Wittfogel, an FAO food processing expert, notes the Burmese "liking for rather strong and at times even 'stale' flavors of fish products."[4] He goes on to state, "When non-deodorized fish flour, with the smell and taste of fish meals processed from sharks for animal feeding, was offered to the various staff members of different races at the Union of Burma Applied Research Institute in Rangoon,[5] it was declared 'delicious' and was even preferred to the various experimental dishes of Burmese soups and fried vegetables, prepared with shark powder and dried prawns." Pranger and Ross did not make test studies that might have provided an economic justification for pursuit of their project, and the project ended with their departure.

The work of Pranger and Ross did, however, kindle interest in the idea of using fish meal as a substitute for dried prawn (a common and not inexpensive item of the Burmese diet). Early in 1959, attempts to make a powder from dried bilsa (a fish with a higher oil content than the sandfish) and shark were pursued in the Food Technology Section of UBARI. The bilsa product was considered to be unsatisfactory, because of the high oil content, and the costs of both products were estimated to be as high as, or higher than, the cost of the product they were intended to replace. On this basis the UBARI project did not go beyond the stage of laboratory investigation.[6,7]

Included in Wittfogel's report is a detailed proposal for the construction of three small reduction plants, each to process 12 tons of raw material daily to make fish meal for human consumption. These plants were to be located in the Irrawady Delta area. The raw material for the proposed plants was to be the "middling" fish, that is, fish under 6 inches in length, that were at that time being added to *ngapi* that was subsequently sold as "pure shrimp" paste. Wittfogel assumed that the fish then caught in the Irrawady Delta (12,000 to 20,000 tons annually) would become available for reduction upon the passage of legislation that would amend the Food and Drug Act of Burma to make such adulteration illegal.

In anticipation of proposed expansion of the Burmese sea fishery Wittfogel also suggested that plans be made to construct a fourth 12-ton reduction plant for food-grade fish meal in Rangoon and to make use of the additional amounts of trash fish expected to be in the trawl harvest. It is interesting to note that he also proposed that a separate five-ton reduction plant be built "to process the offals from the filleted big fish and from the extremely bony fish" into "crude fish meal for animal feeding," and that "both plants should be run complementary to each other. . . ." (This approach contrasts sharply with the Société Nationale de Farine Alimentaire de Poisson (SONAFAP) approach in Morocco but is similar to the one used by the Institute for Fisheries Development (IFOP) in Chile.)

It is not known whether Wittofgel's report further influenced the government or private interests to embark on the commercial production of this type of FPC. However, at the August 1960 meeting of the PAG, FAO reported that 50,000 pounds of fish flour had been produced and marketed. The product was considered by FAO to be "highly acceptable" and they reported that the firm intended to "seek UNICEF assistance."[8] No further development, however, is known.

As far as current activity is concerned, Dr. Maung Maung Gale of the Central Research Organization (CRO) reported recently, "We are still interested but only in an academic way, it seems. A few months ago the Chemistry Department of the Rangoon Arts and Science University produced an M. Sci. thesis entitled, 'Preliminary Investigations on the Production of Fish Protein Concentrate (FPC) from Burmese Sea Fish.' An Abstract of that thesis shows that a solvent-extraction process was investigated. Analytical results were encouraging, with the product judged to be 'a very suitable protein supplement to diets based on stable cereals.' "[9]

References

1. Most of the information in this section was supplied by Dr. Maung Maung Gale of the Central Research Organization (CRO) in Rangoon. The authors would like to express their appreciation to Dr. Gale for his kindness and co-

operation. Without his assistance very little information about the history of interest in FPC in Burma would have been available.

2. Will Pranger, and Mary Ross, "Preparation of fish flour as conducted in the test kitchen of the Nutrition Project Burma," (August 15, 1956). (Report supplied by CRO.)

3. Quoted in H. Wittfogel, "The development of the food industry in Burma. Proposal for the establishment of a fish flour industry for human consumption in Burma, 1960. Dr. Wittfogel was in Burma from October 1959 to November 1960 as an FAO food processing expert. His report was supplied by the CRO.

4. *Ibid.*

5. The Union of Burma Applied Research Institute (UBARI) was the former name of the Central Research Organization (CRO).

6. Maung Maung Gale, personal communication, 1975.

7. S. Ghosh, "Dehydrated bilsa fish," Internal report, Food Technology Section, Applied Chemistry Research Department, UBARI, ca. 1959.

8. Nutrition Document R. 8/Add. 10. WHO/FAO UNICEF PAG, August 1960 meeting, Washington, D.C., p. 5.

9. Ma Htwe Yin, "Preliminary investigations on the production of fish protein concentrate (FPC) from Burmese sea fish," Abstract of M. Sci. thesis, Chemistry Department, Rangoon Arts and Science University, 1975.

COLOMBIA

As far as can be determined, FPC experience in Colombia has been confined to acceptability trials, conducted by the Colombian Institute of Family Welfare, with the collaboration of CARE. The trials used FPC supplied by the U.S. Agency for International Development (AID), from the Experiment and Demonstration Plant (EDP) at Aberdeen, Washington, USA. Two studies were carried out: the first, in 1973 and 1974 with 2857 preschool children and 484 lactating mothers,[1] and the second, in 1974 with 9305 school children.[2] The latter study included a comparison of bread supplemented with FPC, CSM, and DSM. Both trials showed that the products tested were uniformly accepted by the test groups.

No further experiments, nor production, are contemplated.[3,4]

References

1. Franz Pardo, Eberth Betancourth, Reinaldo Grueso, and Maria Dolores Gomez, "Acceptabilidad de un concentrado de proteina de pescado (CPP)

en lactantes y pre-escolares," Publication of the Instituto Colombiano de Bienestar Familiar, T.R.I.–58, Bogota, 1975.

2. Franz Pardo, Maria Dolores Gomez, Ebarth Betancourth, and Reinaldo Grueso, "Acceptabilidad de un concentrado de proteina de pescado en escolares," Publication of the Instituto Colombiano de Bienestar Familiar, T.R.I.–59, Bogota, 1975.

3. Guillermo V. Varela, (Jefe, Div. de Poblacion y Nutricion, Departamento Nacional de Planeacion), personal communication, December 18, 1975.

4. Gustavo, Henriques Suarez, (Jefe, Seccion Aprovechamiento–Pesca Maritima), personal communication, October 13, 1975.

DENMARK

Starting in 1963, attempts were made to develop a process for the production of a low-fat protein concentrate by solvent extraction of fish meal. The experiments were conducted at the Technological Laboratory of the Ministry of Fisheries (Fiskeri-Ministeriets Forsogslaboratorium) in Copenhagen, in collaboration with Atlas Ltd., a large manufacturer of fish meal equipment. In some instances support was also supplied by DANIDA, the Danish International Development Agency. The primary purpose was to produce a product with fat content below 0.5 percent for use in pig rations. This approach was not seriously considered for production of FPC for human consumption because of the added cost of milling that would be necessary to reduce the grittiness of the product.[1]

Most Danish effort on FPC has been directed toward development of two different products: a roller-dried minced fish flesh product and a water-soluble hydrolysate. Experiments with the former have been underway since 1970 at the Ministry of Fisheries Technological Laboratory using homogenized flesh from skinless fillets of cod, haddock, garfish, dab, sand eel, flounder, herring, and dogfish. The product has a texture considered desirable when it is minced with other foods and has been successfully incorporated into a variety of dishes, including potato soup, kroepoek (an Indonesian fried snack), graham bread, fish soufflé, and rice. Development continues in an effort to refine processing conditions and reduce costs.[2,3]

The second of the two products mentioned, the fish hydrolysate, is an outgrowth of the "fish silage" process used for many years by Danish fish producers, that involves acid hydrolysis of "trash" fish on board ship. The hydrolysate for human consumption is essentially a product of autolysis, with pH lowered by addition of formic acid. Attempts to remove the formic acid after hydrolysis have involved reverse osmosis, solvent extraction, spray drying, and vacuum distillation. In a 1973 report of these efforts, however, not much hope was held out for the eventual success of this approach to FPC production.[4,5]

References

1. Lars Herborg, Letter from Fiskeriministeriets Forsogslaboratorium to Mr. Erling Hulgaard, Fisheries Attache, Royal Danish Consulate General, New York, N.Y., September 24, 1975.

2. See note 1.

3. Lars Herborg, F. Wilien, and A. Bruun. "Roller-dried fish protein," paper presented at the FAO Technical Conference on Fishery Products. Tokyo, April 11-12, 1973 (FAO Document FII: FP/73/E-26, May 1973).

4. Jorgen Jensen and Preban Ikkala, "A fish protein hydrolysate and fish sauce: A statement of the work carried out in Denmark," Presented to the IAFMM Scientific Committee Meeting, April 1973.

5. Information for this section was also supplied by Mr. Bjorn Olsen, Head of Division, Danish International Development Agency, Copenhagen, in a letter of August 14, 1975.

FRANCE

Apart from one article[1] and two French patents[2,3] dealing with FPC, only the work of P. Tord, referred to in chapter 7 on FPC developments in Morocco, are evidence of interest in this subject in France.

References

1. C. Cheftel, "Continuous enzymatic solubilization of fish protein concentrate: Enzyme recycling studies," *Ann. Tech. Agric.*, 21:3 (1972), pp. 423-433.

2. C. M. Libenson, and I. Piroskh, "Odorless and Tasteless Fish Protein Concentrate," French Patent 1,502,113 (cl. A 23) No. 18, 1967.

3. C. Rouchy, "Solvent and Method for Extraction of Undesirable Materials from Natural Protein Products," French Patent 2,225,098 (Cl. A 23j), 1974.

GERMANY

During World War II, a product known as "Wiking Eiweiss" (Viking Egg-white) was manufactured in Germany. This soluble product was used as a substitute for some of the protein foods that grew so scarce in Germany toward the end of the war. A description of the product, its preparation, and use was provided by a British Intelligence team at the end of the war. "The raw material used is the flesh of white fish such as cod. During the war the raw material used was mainly 'stockfish' brought from Norway. This gave an inferior product—yellowish instead of pure white, and of a distincly salt fishy aroma—but one that nevertheless was successfully used in soup kitchens."[1] The minced fish flesh was treated with acetic acid to remove some undesirable components and then extracted with trichlorethylene to lower the fat content. This product was then subjected to alkaline hydrolysis, neutralized, and spray-dried to a white soluble powder. In addition to its use in soup kitchens, the product was sold on the retail market and to the food industry as an egg-white substitute. Production of Wiking Eiweiss was halted in 1949.[2]

After a gap of several years, interest in FPC production began to revive. In the late 1950s, a process was developed that reputedly involved "the use of electrophoresis and electrostriction to remove the cellular fluids from the tissue of fish and fish offal." It was reported to obviate the need for heating, to retain a variety of vital nutrients otherwise destroyed, and to reduce the extraction costs by 30 percent.[3] This process, though patented in Germany, was never commercially developed nor, indeed, have the claims been verified by independent investigation.

A variety of other processes for preparation of FPC were proposed in the ensuing years, including the methods of Hartmann (for extraction of herring meal) and Vogel (for treatment of fish and fish meal by a combination of acidification and extraction with ethyl alcohol). In any event, no commercial FPC manufacture exists or is now contemplated in Germany, nor has there been any use of the product since 1949.

References

1. "Certain aspects of the German fishing industry," Final Report No. 4R3, Item No. 22, Target No. C22/328, British Intelligence Objectives Subcommittee.

2. Klaus Wiendieck, personal communication, 1976, Dr. Wiendieck is Counselor, Scientific and Technological Affairs, of the Embassy of the Federal Republic of Germany.

3. Foreign Service Dispatch D–59, from the American Consul General, Bremen, to Department of State, December 30, 1958. The Process was developed by Heinz Doevenspeck.

GHANA

Attempts to produce FPC were initiated in 1950 by F. R. Johnson, Chief Fisheries Officer of the Ghanaian Fisheries Department.[1] Information of FPC-related activities in Ghana is confined to a single report. At the August 1960 meeting of the PAG, FAO reported that in the period since the previous

February, "an interesting fact has been revealed from the examination of a fish flour produced at the Osu Fisheries Station, Accra, Ghana. The fish flour (FAO No. 210) made from *Sardinella* and containing the fish solubles had a fat content of 7.7 percent and would be expected to become extremely rancid during storage. After storage at room temperature for nine months it . . . had no detectable rancid flavor."[2] The report also indicates that analysis of the sample revealed the presence of enough natural antioxidant to prevent oxidation of the fats.

Subsequently FAO reported, "nondeodorized fish flour has been successfully marketed in the Belgian Congo and Ghana, at first at a specially reduced price and later at a commercial price. Apparently there are prospects of increasing demand in these countries."[3]

References

1. From a letter by K. B. Impraim, for Head of Mission, Embassy of Ghana in Washington, D.C., dated October 1, 1976.

2. Nutrition Document R. 8/Add. 10, WHO/FAO/UNICEF PAG, August 1960 Meeting, Washington, D.C., p. 7.

3. FAO Nutrition Division, "1961 Edible fish flour—Development and potentialities. February 1961, Presented at FAO International Meeting on Fish Meal, March 20–29, 1961, Document No. IFIME/W/4.

GUATEMALA

Use of FPC in Guatemala has been confined to investigating its use as a protein supplement to improve the protein quality, and content, of a variety of foods based on cereal grains. This work has been carried out principally by R. Bressani and coworkers at INCAP, the Instituto de Nutrición de Centro America y Panama. Early experiments were conducted with VioBin FPC and were concerned with supplementation of lime-treated corn flour used for making tortillas.[1,2] Subsequent investigations were aimed at supplementation of rice.[3,4]

References

1. R. Bressani, "Enrichment of lime-treated corn flour with deodorized fish flour (fish protein concentrate)," FAO International Conference On Fish In Nutrition, September 19–27, 1961, Washington, D.C., in *Fish in Nutrition,* Fishing News [Books] Ltd., (London: 1962). pp. 266–270.

2. R. Bressani and E. Marenco, "The Enrichment of lime-treated corn flour

with proteins, lysine and tryptophan, and vitamins," *J. Agr. Food Chem.* 11:6 (1963), pp. 517-522.

3. L. G. Elias, R. Jarquin, R. Bressani, and C. Albertazzi, "Suplementación del arroz con concentrados proteicos," *Arch. Latinoamer. Nutr.*, 18:1 (1968), pp. 27-38.

4. Julio Asensio Wunderlich, personal communication, 1975. Information on the current status of FPC in Guatemala was kindly furnished by His Excellency Julio Asensio Wunderlich, Ambassador from Guatemala to the United States.

HUNGARY

The record of FPC activities in Hungary is confined to a single report on the preparation of a protein product from "mass" fish, used as a filler for canned meat products and sausages.[1]

References

1. R. Lorinez, "The manufacture of protein from home-produced, so-called 'mass' fish," *Elelmez. Ipar, Budapest,* 5 (1951), pp. 72-75.

ICELAND

Experiments with fish flour are recorded as having been carried out in Iceland in 1938. At that time a flour was made by grinding vacuum-dried skinned cod fillets, with no attempt made to remove the fatty material or to deodorize it. The flour was incorporated in various soups and breads, reportedly "successfully."[1]

Apparently this line of investigation soon ended, and the next attempt at FPC experimentation occurred after World War II. As part of its Marshall Plan aid, the United States gave Iceland an "HTM" plant, that is, a plant to dehydrate ground fish in a heat transfer medium—in this case, fish oil—that would keep the particles separated as the water was driven out of them by heat and removed from the system by a partial vacuum. The solid, dry fish particles were then separated from the oil, and defatted by solvent extraction. Although they were carried out by Faxi-Ltd., a large fish meal producer, these experiments were intended to investigate the preparation of FPC. Faxi also experimented with various other methods of drying fish and then extracting the dried product with hexane. Although the exact dates are difficult to determine, it seems likely that all these experiments took place between 1952 and 1954, after which time nothing more seems to have been done until the early 1960s when experiments with the HTM method were resumed for a period.[2]

The net result of this very early and continued interest seems to be nothing; there is no record of any commercial manufacture of FPC, or further experimentation, in Iceland.

References

1. G. Hannesson, "Fish flour (fish protein concentrate)—Technological developments in Iceland," FAO International Conference on Fish in Nutrition, September 19-27, 1961, Washington, D.C., in *Fish in Nutrition* (London: Fishing News [Books] Ltd., 1962), pp. 416-418.

2. "Fish Protein Concentrate (FPC) Program Review and Plan, Part I," Report Prepared for National Marine Fisheries Service, NOAA, Washington, D.C., by Northrup Service, Inc., p. G-10.

INDIA

India has a longer history of concerted attempts to prepare and use FPC than the five countries studies in detail in earlier chapters. It is a history of successes and failures in processing experiments and acceptability tests, which ends where all other such histories have ended, with the exception of Sweden and Poland; that is, no current production or use of FPC is taking place in India.

One of the earliest published accounts of Indian interest appeared in 1955 when a process was described that produced a soluble FPC made from shark and ray, as examples of "waste" fish.[1] The cream-colored water-soluble powder that resulted from alkaline hydrolysis of extracted fish flesh was not only recommended as a food for the malnourished and for ulcer patients with a need for bland, water-soluble protein, it was also described as useful in the plastics, paint, and rayon fiber industries. Subsequent concern with FPC use in India, however, was directed exclusively toward its use as a food supplement.

By 1960, acceptability studies using a sample of Astra FPC, supplied by FAO, were under way in India[2] and soon experimentation with processes, raw materials, and studies of nutritive value of the product became quite widespread. A method for preparing "fish flour for human consumption" by fermenting minced shark flesh in defatted buttermilk was reported in 1961,[3] and it was soon followed by reports of an ethanol extraction process for making FPC from the oil sardine (*Clupea longiceps*).[4,5,6] A series of well-conducted nutritive value and shelf-life studies was conducted in 1962, testing a variety of dietary combinations in rat-feeding studies[7–12] before experiments were begun in 1962 to determine the efficacy of FPC as a protein suppoement in the diets of children. These experiments with the FPC made from *C. Longiceps* by ethanol extraction, gave promising results.[13,14,15] About this time, experiments with preserving fish in ammonia solutions, prior to extraction and additional processing into FPC, were reported[16] but no significant further work on this approach seems to have been performed.

The general problems facing the introduction of new (protein-rich) foods and food supplements, such as FPC, began to be discussed in 1963. In discussing these difficulties at the 1963 PAG meeting in Geneva, A. Sreenivasan gave a summary of the status of FPC work in India:

> Although considerable interest has been shown in the possibilities for utilization of the vast fish resources of the sea, not much progress has followed initial efforts at obtaining fish flour preparations, relatively free from odour. A process has been developed at Mysore [Central Food Technological Research Institute (CFTRI)] for obtaining odourless fish flour from oil sardine after cooking pressing for recovery of oil and successive extractions of the dried meal with ethanol. Similar perparations are obtainable from shark (*Carcharias laticaudus*) and from Bombay duck (*Harpodon nehereus*),[17–19] all of which are had in abundance in seasons of glut in various coastal regions of the country. More recently, it has been demonstrated that steeping, soon after catch, in low concentrations of ammonia could well preserve the fish for extended periods of time so that the material can straightaway be dried and processed for fish flour production. The various products have been nutritionally evaluated and a number of fish flour-enriched preparations have also been prepared and tested for protein quality. Pilot-scale production plants are envisaged and it is expected that ere long commercial fish flour production by feasible procedures and from different fishes will become possible.[20]

Despite this optimistic outlook, commercialization of FPC production never materialized in India, although a variety of investigations continued throughout the 1960s and into the early 1970s, experimenting with various processing modifications and raw materials.[21–27] Diminishing interest in India, however, seems to be less related to the economic problems that have plagued FPC use in other countries than to a series of discouraging reports concerning its usefulness in treating clinical protein malnutrition. Starting in 1964[28] these reports continued for several years, essentially contradicting the favorable results obtained earlier[29] and describing the failure of FPC feeding tests to provide as good indices of recovery from protein deficiencies as dried skim milk,[30] or as vegetable protein supplements.[31]

The shift in attitude toward the usefulness of FPC in the treatment of severe protein and protein-calorie malnutrition in India is particularly curious in that it contradicts not only earlier experience in India but also earlier, and subsequent, experience with FPC in other countries. It is worth noting, therefore, that the shift coincides with the use of FPC made by extraction with ethylene dichloride, and the discouraging results obtained with this product. These results are in accord with the lower PER values that have consistently been found for FPC's made by this process. Low PER value was one of the main reasons for U.S. AID's cancellation of its contract with Alpine Marine Protein Industries, whose FPC was made by the ethylene dichloride process. (See chapter 2.)

Currently, there is little, if any, interest in FPC in India. In response to a recent inquiry about the status quo, the following comment, received from the Director General of the Indian Council of Medical Research accounts for this situation.

> Studies have been carried out both at the National Institute of Nutrition, Hyderabad (India) as well as the Nutrition Unit of the Christian Medical College and Hospital, Vellore, Tamilnadu (India) regarding acceptance of fish flour preparations and their effectiveness in rehabilitating children suffering from severe forms of protein-calorie malnutrition . . . the main finding with regard to fish protein concentrates was that its [*sic*] acceptability was not good and also its biological effectiveness was not quite satisfactory.[32]

References

1. G. B. Mohanty, and A. B. Roy, "Hydrolyzed fish protein from the flesh of waste fish," *Science,* 121 (1955) pp. 41-42.

2. Nutrition Document R.8/Add. 10, WHO/FAO/UNICEF/PAG, August 1960 Meeting, Washington, D.C.

3. FAO Indo-Pacific Fisheries Council, Section A. January 6-23, 1961.

4. N. L. Lahiry, M. N. Moorjani, K. Visweswariah, S. R. Shurpalekar, M. Swaminathan, A. Sreenivasan, and V. Subrahmanyan, "Preparation of edible fish flour from oil sardine (*Clupea longiceps*), *Food Science,* (Mysore), 11:2 (1962), pp. 37-39.

5. S. B. Kadkol, and N. L. Lahiry, "Amino acid composition of defatted fish flour from oil sardine (Clupea longiceps)," *J. Sci. Ind. Res.,* 21D (1962), pp. 387-388.

6. M. N. Moorjani, N. L. Lahiry, R. B. Nair, A. N. Upadhye, and S. V. Rao, "Nutritional value of fish flour, I: Effect of storage of sardine meal prior to its extraction with ethanol," *Food Technology,* 19 (1965), pp. 212-215.

7. S. R. Shurpalekar, N. L. Lahiry, M. N. Moorjani, M. Swaminathan, A. Sreenivasan, and V. Subrahmanyan, "Chemical composition and shelf-life of a protein food based on low-fat peanut flour," *Food Science* (Mysore) 11:2 (1962), pp. 39-41.

8. S. R. Shurpalekar, M. N. Moorjani, N. L. Lahiry, K. Indiramma, M. Swaminathan, A. Sreenivasan, and V. Subrahmanyan, "Supplementary value of

proteins of fish flour to those of peanut flour and the protein efficiency ratio of a protein food containing peanuts, Bengal gram, and fish flours," *Food Science*, pp. 42-44.

9. S. R. Shurpalekar, A. A. Joseph, N. L. Lahiry, M. N. Moorjani, A. N. Sankaran, M. Swaminathan, A. Sreenivasan, and V. Subrahmanyan, "Supplementary value of fish flour and a protein food containing low fat peanut flour, Bengal gram flour, and fish flour to poor rice diet," *Food Science*, pp. 45-48.

10. S. R. Shurpalekar, A. A. Joseph, M. N. Moorjani, N. L. Lahiry, K. Indiramma, M. Swaminathan, A. Sreenivasan, and V. Subrahmanyan, "Supplementary value of fish flour fortified with vitamins to poor Indian diets based on different cereals and millets," *Food Science*, pp. 49-51.

11. S. R. Shurpalekar, A. P. Jayaraj, M. N. Moorjani, N. L. Lahiry, A. N. Sankaran, M. Swaminathan, A. Sreenivasan, and V. Subrahmanyan, "Supplementary value of fish flour and a protein food containing low-fat peanut flour, Bengal gram flour, and fish flour to a corn-tapioca diet," *Food Science*, pp. 52-56.

12. S. R. Shurpalekar, A. A. Joseph, N. L. Lahiry, M. N. Moorjani, M. Swaminathan, N. Nataraja, A. Sreenivasan, and V. Subrahmanyan, "Relative value of a protein food containing fish flour, peanut flour, and Bengal gram flour as compared with skim milk powder in meeting the protein requirements of protein-depleted rats," *Food Science*, pp. 57-61.

13. T. R. Doraiswamy, S. R. Shurpalekar, M. N. Moorjani, N. L. Lahiry, A.N. Sankaran, M. Swaminathan, A. Sreenivasan, and V. Subrahmanyan, "The effect of a supplementary protein food containing fish flour, groundnut flour and Bengal gram flour and fortified with vitamins on the growth and nutritional status of children," *Indian Journal of Pediatrics*, 30:187 (1963), pp. 266-271.

14. S. R. Shurpalekar, V. A. Daniel, T. R. Doraiswamy, N. L. Lahiry, M. N. Moorjani, and M. Swaminathan, "The effect of a supplementary protein food containing fish flour (from oil sardine) on the metabolism of nitrogen, calcium, and phosphorus in children," *Indian Journal*, pp. 272-275.

15. A. Sreenivasan, "Fish flour and its utilization," *Nutrition* Document R.8/ Add.14, WHO/FAO/UNICEF PAG, August 1963 Meeting, Geneva.

16. See note 15.

17. G. D. Revankar, V. S. Khabade, and S. V. S. Rao, "Urea-free fish protein concentrate from shark meat," *Res. Ind.* (New Delhi), 10 (1965), pp. 346-367.

18. D. P. Sen, and T. S. Satyanarayana Rao, "Deodorization of fish protein

concentrate from Bombay duck (*Harpoden nehereus*)," *Journal of Food Science Technology* (Mysore), 3:1 (1966), pp. 27–28.

19. D. P. Sen, T. S. Satyanarayana Rao, and N. L. Lahiry, "Defatting and deodorization of fish protein concentrate from *Harpodon nehereus*," *Journal of Food Science Technology*, 31:3 (1966), pp. 344–350.

20. A. Sreenivasan, "Problems and experience in the marketing of protein rich foods in India, Nutrition Document R. 10/Add. 54, WHO/FAO/UNICEF PAG, August 1963 Meeting, Geneva. This optimistic outlook continues to be expressed periodically. In 1970, for example, the Information Service of the Embassy of India in Washington, D.C., in its *India News*, published the following item: "A completely odourless fish flour with almost 100 percent protein content has been developed from trash fish by the Central Institute of Fisheries in Cochin. The Director of the Institute, Mr. V. K. Pillai, said the Institute would soon have a pilot plant for the production of this flour." (*India News*, IX:2 [April 3, 1970].)

21. M. N. Moorjani, A. N. Upadhye, N. L. Lahiry, H. A. B. Parpia, M. E. Dumm, and S. M. Pereira, " 'Fricola' a fish-enriched farinaceous product. Feeding tests with kwashiorkor children," *World Fishery Abstracts*, 15:4 (1964), pp. 41–42. (paper presented at FAO Symposium on the Significance of Fundamental Research in the Utilization of Fish, May 26–30, 1964, Husum, Germany.)

22. M. A. Krishnaswamy, S. B. Kadkol, and G. D. Revankar, "Nutritional evaluation of an enriched product from fish," *Canadian Journal of Biochemistry*, 43:11 (1965), pp. 1879–1883.

23. B. L. M. Desai, M. K. Hariharan, A. P. Jayaraj, S. V. Rao, and M. Swaminathan, "Studies on the protein efficiency ratio of fish flour (from Bombay duck [a lizard]) and a protein food based on fish, groundnut and Bengal gram flours and their supplementary value to low protein diets," *Journal of Nutr. Diet*, 5:1 (1968), pp. 45–51.

24. M. N. Moorjani, R. B. Nair, and N. L. Lahiry, "Quality of fish protein concentrate prepared by direct extraction of fish with various solvents," *Food Technology*, 22:12 (1968), pp. 1557–1561.

25. D. P. Sen, T. S. Sayanarayana Rao, S. B. Kadkol, M. A. Krishnaswamy, S. Venkata Rao, and N. L. Lahiry, "Fish protein concentrate from Bombay duck (*Harpoden nehereus*) fish: Effect of processing variables on the nutritional and organoleptic qualities," *Food Technology*, 23:5 (1969), pp. 85–90.

26. S. Pushpama, B. B. Parrish, and C. W. Deyoe, "Improving protein quality of millet-M, sorghum-m and maize-M diets by supplementation," *Nutr. Rep. Int.*, 5:2 (1972), pp. 93–100.

27. S. M. Pereira, T. Isaac, B. Teivarson, and M. E. Dumm, "Processed fish protein in the treatment of kwashiorkor," *Indian Journal of Medical Research,* 53 (1965), pp. 651-658. This report is also curious, in that it contradicts an earlier report by some of the same authors describing favorable results in kwashiorkor treatment with "Fricola" (see note 20).

28. See note 27.

29. See notes 13, 14, and 15.

30. S. G. Srikantia, and C. Gopalan, "Fish protein concentrates in the treatment of Kwashiorkor," Nutrition Document R.8/Add. 21. PAG (WHO/FAO/UNICEF), July 1965 Meeting, Rome.

S. G. Srikantia, and C. Gopalan, "Fish protein concentrates in the treatment of kwashiorkor," *American Journal of Clinical Nutrition,* 18:1 (1966), pp. 34–37.

31. S. M. Pereira, A. Begum, R. Sundaraj, and M. E. Dumm, "Effect of dietary protein on serum amino acids," *American Journal of Clinical Nutrition,* 21:2 (1968), pp. 167–172.

32. C. Gopalan, personal communication, September 4, 1975.

INDONESIA

There is not much information available about the history of FPC use in Indonesia. We do know, however, that attempts have been made to improve the nutritional status of protein deficient segments of the population through the use of a variety of FPCs.

One of the earliest developments occurred in 1958,[1] when preliminary investigations by the Nutrition Board, Djakarta, indicated that the biological value of FPC was comparable to that of milk protein. A subsequent report of these investigations appeared in the worldwide summary presented to the 1960 meeting of the PAG, but contained no details.[2] In 1962, the Department of Child Health of the University of Indonesia, Jakarta, reported the "curing effects" of FPC in treating kwashiorkor in children.[3]

An interesting series of experiments involved the use of *tempeh,* a common food in Indonesia prepared from soy meal that is fermented by a fungal inoculum. The product of this fermentation, although it contained all the oxidizable fats from the soy beans, was known to be stable and it was found that the fermentation process produces an antioxidant that accounts for this stability. Thus, experiments were initiated to determine if *tempeh* could be mixed with nondefatted fish flour (sometimes called "non-deodorized"), more commonly referred to as Type C FPC, to see if the antioxidant properties of the *tempeh* would reduce or prevent rancidity of the fish oils remaining in the Type C FPC. The results showed that the FPC-tempeh mixtures were indeed stable, but we do not know if this product has subsequently been produced or used in Indonesia.

References

1. L. M. Rakjat, "Fish Flour," *Indo-Pacific Fisheries Council Proceedings*, 7:2/3 (1958), pp. 97-99.

2. "World-wide developments in fish flour," Nutrition Document R.8/Add. 10, WHO/FAO/UNICEF/PAG, August 1960 Meeting, Washington, D.C.

3. I. M. Sutedja, S. I. Wenarso, and S. Sanjota, "Fish flour trials on kwashiorkor," *Pediat. Indon.* 2:87 (1962).

ISRAEL

Interest in FPC in Israel never went beyond the stage of preliminary laboratory investigations. In 1967, the U.S. Bureau of Commercial Fisheries received an application for a research grant from the Sea Fisheries Research Station in Haifa to finance a joint project with the National Physical Laboratory of Israel to determine the technical feasibility of small to moderately sized plants for making FPC by the BCF process. The project was intended to determine the feasibility of plants "suitable for local waste fish products, for possible use on board of large trawlers, as well as for utilizing marine resources in underdeveloped areas where no proper transport or storage facilities are available." The research grant was to have been financed by local currency, held by the United States under PL 480. The project was never approved by the BCF.

The last report of FPC-related activity dates from 1969 when the suitability of various fishes from the Mediterranean and Red Seas as raw materials for FPC was investigated.[1] Since then, no other activity related to FPC has taken place in Israel.[2]

References

1. A. Herzberg and R. Pasteur, "Proximate composition of commercial fishes from the Mediterranean Sea and the Red Sea," *Fish Ind. Res.* 5:2 (1969), pp. 39-65.

2. Ammon Barak, personal communication, October 17, 1975. Dr. Barak is Head, Life Sciences Division, of the National Council for Research and Development.

JAPAN

Japan is one country where the conventionally accepted meaning of fish protein concentrate is sometimes irrelevant since fish is eaten in many forms in which the protein concentration is greater than in the original fish. Consequently it is useful to analyze Japan's interest in FPC in two categories: conventional fish protein concentrates, meant as dietary supplements; and partially or completely hydrolyzed products meant for use in the variety of sausagelike foods consumed in Japan.

Relevant to the latter category are two series of papers published by the Tokai Regional Fisheries Research Laboratory, under the general heading, "Utilization of Fish Protein" (starting in 1941), and "Liquefied Fish Protein" (starting in 1965). The first series seems to concentrate on acid hydrolysis of fish and other proteins, and "extraction" of proteins with alkali.[1] The second series of papers is concerned primarily with enzymatic hydrolysis of fish proteins designed to make a greater proportion of marine animal protein water-soluble as a foodstuff of high nutritive value and digestibility for the human diet.[2]

In the category of conventional fish protein concentrates, studies were made of fish flour from codfish in the 1940s[3] and a variety of studies have been made of processes to make an edible product from fish meal. These have varied from enzymatic hydrolysis of fish meal-starch combinations by fungal fermentation[4] to solvent extraction of fish with diethylamine and benzene[5] or a mixture of acetone and benzene.[6] Autolysis of ground fresh fish to form a water-soluble FPC has also been attempted. This product, after its nutritive value was determined to be similar to that of isolated fish protein, was mixed with wheat flour, stabilized vegetable oil, calcium carbonate, yeast, and cane sugar to form a breast milk substitute that was shown to be effective in nursing infants a few weeks old.[7]

Aside from a general review of FPC manufacturing methods that appeared in 1968,[8] no further work with FPC seems to have been undertaken in Japan.

References

1. Examples are: T. Takahashi, "Utilization of fish protein, I: Hydrolysis of codfish meal by hydrochloric acid," *Journal of the Society of Chem. Ind. Japan* 44 (1941), pp. 712–717; and T. Takahashi, "Utilization of fish protein, III: Relation between the state of protein and hydrolysis," *Journal of the Society of Chem. Ind. Japan* 45 (1942), pp. 872–875.

2. Examples are: H. Higashi, S. Murayama, T. Onishi, S. Iseki, and T. Watanabe, "Studies on 'liquefied fish protein,' I: Nutritive value of 'liquefied fish

protein,' " *Bulletin of Tokai Reg. Fish. Res. Lab.*, 43 (1965), pp. 77-86; and T. Onishi, and H. Higashi, "Studies on 'liquefied fish protein,' " *Bulletin of Tokai Reg. Fish. Res. Lab.*, 55 (1968), pp. 225-235.

3. O. Kitayama, "Biochemistry of theragra from the Bering Strait, II: Chemistry of fish flour for food," *Journal of the Agricultural Chemistry Society Japan*, 20 (1944), pp. 566-568.

4. H. Higashi, T. Nitta, K. Nagakura, and S. Umemoro, "Studies on the utilization of fish meal as food," *Bulletin of the Japanese Soc. Sci. Fish*, 17:6 (1951), pp. 147-156.

5. K. Suzuki and K. Saruya, (Tokyo Institute of Fisheries), "Studies on the fish oil-solvent-water system in reference to the manufacture of defatted fish meal," *Nippon Suisan Gakkaishi*, (1964), pp. 37-41.

6. K. Suzuki and K. Saruya, "Preparation of fatless fish meal by solvent extraction," *Nippon Suisan Gakkaishi*, 30:7 (1964).

7. T. Tomiyama, ZESZ. PROBL. POSTEPOW. NAUK. ROLN. 80 (1968), pp. 385-398.

8. S. Murayama, "Fish protein concentrate," *Nippon Suisan Gakkaishi*, 35:5 (1969), pp. 479-486.

KOREA

As is explained in the chapters on Morocco and Chile, Korea was one of the countries surveyed by an AID team as a potential site for the AID FPC pilot project. Korea was not selected, and indeed the Korean experience with FPC has been minimal.

In 1960, experiments were made with a fish powder supplement to a rice diet[1] and in 1961 with a "fish flour" made by sun-drying boiled fish that was then stored in brown paper packages before being ground for use. A sample of this material was submitted to UNICEF which reported on the experiments at the 1962 PAG meeting.[2]

As far as can be ascertained, the only other FPC experiments in Korea were performed with samples of VioBin FPC. The experiments were aimed at testing FPC as a weaning food, and the best that can be said is that the results were mixed.[3] Since then there seems to have been no further activity in this field.

References

1. J. S. Ju, and O. I. Whang, "Nutritional supplementation of rice diet by fish powder," *New Medical Journal* (Korea), 3:10 (1960), pp. 45-53.

2. "Fish Flour," Nutrition Document R.8/Add. 13, PAG (WHO/FAO/ UNICEF), March 1962 Meeting, Rome, pp. 8-10. The sample was submitted by Dr. Jim Soon Ju of Doo Do Medical College at Seoul.

3. K. Y. Lee, M. S. Shin, M. H. Kim, E. S. Lee, B. K. Cho, and J. A. Lewis, "Effect of protein supplementation by fish flour at weaning period," *Yonsi Medical Journal,* 5 (1964).

MALAYSIA

In 1961, at its January meeting, the FAO Indo-Pacific Fisheries Council Section on Processing reported that fish flour for human consumption was being produced as a by-product of the tuna canning operation of the "newly formed" Malayan Marine Products Ltd. (this was before the formation of Malaysia). The Nutrition Division of the Institute of Medical Research was reported to have analyzed the materials and to have recommended its use as a dietary supplement.

Since that time very little work has been done on FPC in Malaysia. Information on the activities that have taken place was supplied by the Malaysian Agricultural Research and Development Institute (MARDI).[1] Research on fish as food has been confined primarily to preserving it, after landing, until it is delivered to the consumer. The use of fish protein concentrate is only in the form of a fermented fish sauce *(Budu),* the manufacture and distribution of which are in private hands. Fish meal production, for animal feed, is active and is the subject of research and development in which both private industry and the federal and local governments are involved.

Reference

1. Ahmad Zaharudin Idrus, personal communication, December 29, 1975. Dr. Zaharudin is Assistant Director, Agricultural Product Utilization Division, MARDI.

MEXICO

Among the early studies of the use of FPC in treatment of malnourished children were those conducted in Mexico. Starting in the early 1950s, rat feeding studies were conducted to test the value of FPC as a supplement to Mexican diets.[1] After positive results were obtained, feeding studies were conducted with children at the Hospital Infantil in Mexico City.[2] The results of these studies were favorable, and this work continued for a number of years. Its current status is uncertain.

References

1. See, for example: M. H. Guillermo, O. Y. Cravioto, R. O. Cravioto, and F. de M. Figueroa, "Studies of proteins and amino acids in Mexican diets, II: Effect of proteins of tortillas and beans used in common diets of the people of Boxaxni, Mezquital Valley, on the growth of white rats," *Ciencia* (Mexico), 13 (1953), pp. 199–203; R. O. Cravioto, *et al.*, "Fish flour as a protein supplement," *Ciencia* (Mexico), 15:4/5 (1955), pp. 83–88.

2. For example: F. Gomez, R. Ramos-Galvan, J. Cravioto, S. French, and I. Labardini, "Studies on the use of deodorized fish flour in malnutrition, Preliminary report," *Bol. Med. Hosp. Infantil* (Mexico). 15:4 (1958), pp.485–493.

THE NETHERLANDS

Various laboratories in the Netherlands have been involved in FPC research and development for many years, starting, perhaps, with the cooperation of the FAO, by analyzing samples of FPC prepared in other countries. Direct activity in FPC development, however, took place primarily between 1960 and 1970. During this period, two institutions were engaged in FPC research development:

1. the Institute for Fishery Products TNO, of the Organization of Applied Natural Sciences TNO, at Ijmuiden; and
2. the Department of Chemical Technology of the Technical University of Eindoven.

In collaboration with the Netherlands chemical firm Akzo N.V., the Ijmuiden laboratory, under its director, If. J. J. Doesburg, developed a process for preparing fish protein isolate (FPI). The FPI was prepared by alkaline hydrolysis of comminuted fresh fish, followed by acidic precipitation of the proteins and extraction with ethanol. This development effort resulted in a patent application (Netherlands #6914370) by the industrial sponsor, Akzo N.V.

The initial purpose of the FPI project was the development of new ways to utilize unexploited fish species and fishery waste products. Gradually, however, the focus changed to the industrial production of high quality proteins for human consumption from both fish and soybeans.

At the Technical University of Eindhoven extensive research and development on FPC were carried out both in the laboratory and on a pilot plant scale. Under the direction of Prof. H. A. C. Thyssen, a continuous countercurrent process was developed for dehydration of a wet fish slurry with isopropanol vapor. This investigation, a part of the University's research program on selective dehydration principles, was carried out in close collaboration with a private engineering firm, Continental Engineering, N.V. As with the FPI project at Ijmuiden, the project resulted in patent applications (for example, West German Patent No. 1, 955, 941).

Poor prospects for commercialization of both the FPI process and the dehydration process resulted in termination of both projects about 1970. The industrial sponsors of these projects apparently felt that the cost of extraction and drying equipment, and uncertainty about the actual losses of organic solvents, were too great to warrant further investment. In addition, there appeared to be little commercial demand for neutral fish protein powders.

Reference

1. The authors are indebted to J. P. Tripplaar, First Secretary (Agriculture) of the Royal Netherlands Embassy in Washington, D.C., for the information in this summary.

NORWAY

It is not surprising that experiments on FPC processes were taking place in Norway in the 1950s in view of that country's long history of fish meal manufacture. Indeed, at the time of his visit in 1962, while conducting a worldwide survey of FPC development for the BCF, E. R. Pariser thought that the fish meal then made by the Stord Atlas Company, for example, was of sufficiently high quality to be suitable for human consumption and was certainly suitable as a starting material for FPC processing.

Under the sponsorship of the Norwegian government, research aimed at the production of a tasteless and odorless fish flour for human consumption was conducted for many years by the Fisheries Research Institute in Bergen. Experiments with at least two extraction methods were carried out, one utilizing ethyl alcohol alone and the other in a mixture with chloroform.[1] These experiments resulted in an application for a patent, filed in 1960, on an FPC process using a mixture of solvents.[2]

By 1969, the Institute was planning a continuous countercurrent extraction process using isopropyl alcohol and starting with presscake.[3] By 1972, "Type B" FPC was being produced in sufficient quantity to enable Norwegian Church Aid to ship 700 tons of the product to Bangladesh, several tons to Senegal, and another 1000 tons to Bangladesh in 1973.[4,5] Norway continues to supply FPC to the World Food Program.

Concern for the use of FPC for treating protein deficiencies in developing countries remains strong in Norway, with interest in the use of "Type B" FPC as a "tonic" to prevent protein malnutrition.[6]

References

1. "Fish Protein Concentrate (FPC) Program Review and Plan, Part I," Report prepared for National Marine Fisheries Service, NOAA, Washington, D.C., by Northrop Service, Inc., p. G–7.

2. Nutrition Document R.8/Add.10, WHO/FAO/UNICEF/PAG, August 1960 Meeting, Washington, D.C.

3. George K. Parman, Memorandum to Roland A. Finch, Director, National Center for Fish Protein Concentrate, Bureau of Commercial Fisheries, March 10, 1969.

4. Cato Aall, "Report on Acceptability and possible use of fish protein concentrate in the Southern Sudan," National Nutrition Council/FAO Committee of Norway, P.O. Box 8139, Oslo Dep, Oslo, 1972.

5. Cato Aall, "Some long- and short-term aspects in dealing with nutrition problems in Bangladesh," available as Information Paper No. 27, "Nutrition problems in Bangladesh," United Nations Relief Operation Dacca (UNROD). (See entries on Bangladesh and Sudan for details.)

6. Olav Albert Christophersen, letter to Alan Berg, May 29, 1975.

PAKISTAN

As noted in the history of FPC activities in Bangladesh, interest in FPC in Pakistan predates the U.S. FPC effort. At the PAG meeting in August 1960, it was reported that Pakistan was attempting to produce an acceptable fish flour.[1] Also in 1960, there appeared the first published report describing the process developed at the Karachi laboratories of the Pakistan Council for Scientific and Industrial Research (PCSIR) for production of FPC from shark.[2] In this process the urea was removed by mixing the comminuted shark flesh (and cartilage) with soy powder and heating it to 40° C. The remaining steps of the process were quite complicated, involving several extractions—with oil- and lipid-solvents—and alkaline digestions. The product, a creamy white powder, was essentially odorless and tasteless and was quite pleasing in appearance.

In the years between 1960 and 1970, interest in FPC continued unabated, and a variety of experiments were undertaken. These included further testing of FPC's made from shark[3,4,5] and other fish,[6,7] and experiments with various FPC-supplemented foods.[8–11]

In 1968, there was a move within USAID to provide $274,000[12] to Pakistan for the construction of a pilot plant capable of producing about 150 pounds per day of FPC from shark.[13] Early in 1970, AID conducted a survey of the potential for FPC manufacture in Pakistan which pointed out that there were some major obstacles to be overcome before such a plant could be constructed.[14] Two of the most immediate problems were the absence of an engineering process to serve as a basis for the pilot plant design, and the lack of an assured supply of fish. The former could certainly be overcome with time, but the latter seemed serious in that it was apparent that there was no significant surplus of edible fish; any increase in catch could be consumed as fish, that is, in its original form. With respect to shark, since most of the shark was exported, any significant diversion to FPC would mean loss of foreign exchange. Similar questions were raised by a private consulting firm that performed another review for AID a year later.[15] Furthermore, at the time of the first survey (early 1970), the situation was complicated by the position taken by the central government that two FPC plants should be built—one outside of Karachi in West Pakistan, and one in Cox's Bazaar in East Pakistan. As a result of these problems and of the civil war that broke out in 1971, no further action was taken on this project.

Since that time, however, interest in production of FPC has not died down at the Karachi laboratories of the PCSIR. Interest still persists in the establishment of commercial manufacture of FPC, but nothing definite has yet developed, primarily because of the lack of raw material.[16] The Karachi laboratories of the PCSIR are proposing a government investment of 1.3 million rupees in a pilot plant for production of both a suluble FPC (prepared by enzymatic hydrolysis) and conventional (insoluble) FPC prepared from precooked fish.[17] Assisting the PCSIR in its efforts is the Denver Research Institute. The funds have not yet been allocated, and construction of the plant has not started.

References

1. "World-wide developments on fish flour," Nutrition Document R.8/Add. 10. WHO/FAO/UNICEF PAG, August 1960 Meeting, Washington, D.C.

2. S. Abdul Hag and S. Mahdihassan, "Shark meat flour," *Pakistan J. Sci. Ind. Res.*, 3 (1960), pp. 213-215.

3. S. Mahdihassan, "Salvaged proteins for Pakistan," *Current Affairs Bulletin*, 35 (1962), pp. 12-14.

4. S. M. Ail, A. S. Alvi, and M. Hanif, "Protein value of wheat-flour supplemented with shark meat flour," *Pakistan Journal Sceintific Research*, 16:2 (1964), pp. 41-43.

5. S. Maqsood Ali, "Development of new sources of protein rich foods for ameliorating protein deficiency in Pakistan," *Pakistan Journal of Medical Research*, 7:3 (1967), pp. 198-206.

6. H. N. De, "A note on the relative nutritive value and the total and 'available' methionine tryptophan and histidine in some fish flours," *Pakistan J. Sci. Ind. Res.*, 6 (1963), pp. 299-301.

7. A. S. Alvi, M. Hanif, and S. M. Ali, "Net protein utilization of guar *(Cyamopis tetragonobla)* treated in different ways and enhancement of its protein value by supplementation," *Pakistan Journal Scientific Research,* 16:1 (1964), pp. 1-4.

8. A. S. Alvi, M. Hanif, and S. M. Ali, "Evaluation of the protein quality of some fish flours by determination of 'available' lysine and by bacteriological examination," *Sci. Res. Pakistan,* 1 (1964), pp. 123-126.

9. S. M. Ali, M. Hanif, and A. S. Alvi, "A protein food based on a blend of gram flour, guar seed *(Cyamopis tetragonobla)* flour and fish flour," *Pakistan Journal Scientific Research,* 16:3/4 (1964), pp. 103-104.

10. I. A. Shaikh, M. Arshad, M. Y. Ikram-Ul huq, and S. M. Ali, "Improvement of protein value of cotton seed protein isolate with fish flour and skim milk powder," *Pakistan J. Sci. Ind. Res.,* 10:1 (1967), pp. 86-88.

11. M. B. Bhatti, and R. A. Riaz, "Studies on the 'enrichment' of local bread (or roti) in West Pakistan," *Pakistan J. Sci. Res.,* 19:1 (1967), pp. 7-10.

12. USAID, "Non-capital Project Paper (PROP)," 1969. The funds were to be spread out over three fiscal years: $194,200 in FY 69, $58,800 in FY 70 and $21,000 in FY 71.

13. Norman L. Brown, "Survey of potential for fish protein concentrate manufacture in Pakistan—1970," Report to USAID, 1970.

14. See note 13.

15. American Technical Assistance Corporation, Washington, D.C. "A review of nutrition program possibilities in Pakistan," Report to USAID, 1971.

16. Amir Usman, personal communication, 1975. In a letter to the authors dated October 23, 1975, Mr. Usman (Counsellor to the Embassy of Pakistan in Washington, D.C.) explained that manufacture of FPC in Pakistan had not yet taken place "because no surplus fresh fish is available for this purpose."

17. It is worthwhile to note here the same point that was made in the section on Bangladesh, *viz.,* that the staff members of the PCSIR labs have a personal financial interest in the commercial development of any process or product they develop in the laboratory.

PERU

Peru is one of the few countries where there is a long history of both important experimental use of FPC in combating malnutrition and serious industrial interest in commercial production. Despite the convergence of these efforts, no viable commercial manufacture of FPC has developed. Indeed, an objective assessment leads to the conclusion that Peru's most significant contribution to the FPC story probably lies in the pioneering work of Dr. George G.

Graham and his colleagues with the use of FPC in the treatment of marasmus and marasmic kwashiorkor in children and infants.

One of the early reports of Dr. Graham's work at the British-American Hospital in Lima was presented to the FAO Conference on Fish in Nutrition held in Washington in 1961.[1] Dr. Graham reported on the results of using FPC (supplied by VioBin) in the treatment of undernourished infants in one of the earliest well-documented experiments that demonstrated dramatically the high nutritive value of FPC.

Research interest in FPC continued with experimentation at the British-American Hospital,[2,3,4] the Agrarian University at La Molina, the Faculty of Pharmacy and Biochemistry of San Marcos University,[5] and the National University at Lambayeque.[6]

Experiments with FPC had been taking place even earlier, however, according to a report of acceptability studies with FPC enriched bread started in 1959.[7] In any event, by 1962 interest in FPC in Peru had reached the point where FAO, UNICEF, the governments of Peru and the United States, and the Peruvian fish meal industry were all supporting the idea of establishing a plant.

Private industry indicated its willingness to support and participate in this area to the extent that it was described as being so "anxious to be first in this field commercially that there might be competition among the manufacturers to set up and operate the proposed plant." Furthermore, FAO felt that "industry would also consider contributing towards the cost of foreign experts possibly to the extent of 30 percent."[8]

There seemed to be interest on all sides in Peru in establishing an FPC industry, with the exception of the Fisheries Department, which had its own scheme for promoting consumption of salted and dried anchovetta. The Ministry of Health was to take primary responsibility for the FPC program and agreed to allow existing Health Units to serve as focal points for FPC introduction.

Inquiries were sent by FAO to a number of European companies regarding equipment for the planned FPC plant, but in spite of a fair amount of correspondence, the plans for an FAO-sponsored plant never materialized, probably because FAO witnessed the failure of the UNICEF-sponsored FPC plant in Quintero, Chile, at about this time. (See chapter 8.)

Nevertheless, commercial production of FPC was established in at least one plant in Peru. In 1962, a patent was granted to Carlos Varrando Bruera[9] for a process of solvent extraction of comminuted fish that exposed the particles to hot solvent (hexane) vapors under conditions of reduced pressure. The process, similar to the common industrial practice known as "vapor degreasing," was claimed to reduce the lipid concentration of hake (*Merluccius gayi gayi*) to less than 1 percent, and the water concentration to no more than 8 percent, with the use of an amount of solvent equal to 1.5 times the weight of the raw material. The product was licensed by the Ministry of Public Health on June 3, 1964, as "Supplement nutritivo compensatorio" for human consumption. The Varrando product—manufactured by the CIVSA company—was used by the Ministry of Public Health in a bread enrichment campaign starting in 1964 and in a school breakfast program. A precooked food mixture made with this product (about 50 percent FPC) was also marketed.[6] Finally, a large wheat milling and noodle producing firm, Nicolini Hnos. S.A., which had been experimenting with enriched noodles (presumably made with VioBin FPC), began in 1965 to study the possibility of using the Verrando FPC in its products.

The net result of the commercial activity is not clear. In 1969,[10] CIVSA was reported to be producing 180 tons per month, but whether the product is still being manufactured and sold for human consumption today is not certain.

References

1. G. G. Graham, J. M. Baertl, and A. Cordano, "Evaluation of fish flour in the treatment of infantile malnutrition," FAO International Conference on Fish in Nutrition, September 19-27, 1961, Washington, D.C. in *Fish in Nutrition* (London: Fishing News (Books) Ltd., 1962), pp. 271-274.

2. G. G. Graham, A. Cordano, and J. M. Baertl, "Studies in infantile malnutrition II: Effect of protein and calorie intake on weight gain," *Journal Nutrition,* 81 (1963), p. 249.

3. G. G. Graham, J. M. Baertl, and A. Cordano, "Dietary protein quality in infants and children. 1. Evaluation in rapidly growing infants and children of fish protein concentrate alone and in combination with wheat," *American Journal of Diseases of Childhood,* 110 (1965), pp. 248-257.

4. G. G. Graham, J. M. Baertl, and A. Cordano, "Studies in infantile malnutrition. V. The effect of dietary protein source on serum proteins," *American Journal Clinical Nutrition,* 18:1 (1966), pp. 16-19.

5. Gustave Contesse, "Development of fish protein concentrate in Peru. Verrando Process," Nutrition Document R.8/Add.18. WHO/FAO/UNICEF PAG, July 1965 Meeting, Rome, 1965.

6. M. C. Marquez and E. J. Castaneda, "Fishing industry technology. Odorless fish powder," *Bol. Soc. Quim. Peru,* 32:2 (1966), pp. 85-90.

7. FAO News Release, International Meeting on Fish Meal, March 23, 1961.

8. Fred Popper, "Report of a meeting held in the office of Fred Popper, FAO, on 12 January 1962 to discuss his recent trip to Peru and Chile," 1962.

9. US Patent 3,064,018, November 13, 1962. Other patents covering the process are: Peru No. 4958; Germany No. 1, 178, 173; Britain No. 911628;

Canada No. 688755; Brazil No. 68171; Mexico No. 66612; Chile No. 16301; Argentina No. 114655; Columbia No. 10388; Italy No. 603537; France No. 1220169; and Portugal No. 35379.

10. R. H. Phillips, "Peruvian firm making FPC and using it in other foods," *Nat. Fish.*, 50:8 (1969), p. 7-c.

PHILIPPINES

Experiments with FPC in the Philippines date from 1960 when samples of ASTRA FPC were sent by FAO to Dr. A. Bayan, of the Philippines Women's University for acceptability trials.[1] More recently, studies involving the production and evaluation of FPC have taken place at the Department of Fisheries Technology, College of Fisheries, University of the Philippines. With financial support (P65,000 for three years) from the National Research Council of the Philippines, a research project aimed at testing the suitability of local, underutilized species for production of FPC was undertaken in March 1972 by Florian Magno-Orejana, Rogelio O. Juliano, and Carmelita M. Biqueras. Solvent extraction methods were compared for yield, cost, and adaptability to conditions in the Philippines, using ethanol and isopropanol. The products were subjected to organoleptic, microbiological, and chemical evaluation, and the biological value of the protein was determined.

The results of this project formed the basis for a project entitled, "Multidisciplinary Approach in Bridging the Protein Calorie Gap Through FPC (Fish Protein Concentrate)" that is currently underway. This project is being undertaken jointly by

Department of Fisheries Technology, College of Fisheries, University of the Philippines, Diliman, Quezon City;

Institute of Social Work and Community Development, University of the Philippines, Diliman, Quezon City;

Department of Food Science and Nutrition, College of Home Economics, University of the Philippines, Diliman, Quezon City;

Food and Nutrition Research Institute-NSDB, Manila; and

Department of Public Information, Manila.

The project seeks to develop an ideal fish protein concentrate that can be manufactured economically on a large scale for mass feeding of the undernourished segments of the Philippine population. The use of local species of fish which have low consumer demand is given emphasis. The study also seeks to find ways of utilizing FPC by incorporating in the product formulation intended for children.

Funds for the project total to P369,000, P50,000 of which is provided by the University of the Philippines and the balance is coming from the Bread for the World, Federal Republic of Germany.

To date, production of FPC in large quantities is being undertaken at the Department of Fisheries Technology. Distribution of the product in undernourished segments of the country is in progress, this aspect being done by the UP Institute of Social Work and Community Development. Incorporation of FPC at various levels in Kroepoek, a popular snack item of Filipinos which has gained wide acceptability of the consumers has been tried at the UP College of Fisheries; the UP College of Home Economics, on the other hand, has done the same on other snack items like pan de sal, ginger bread and other food items like sausage. The acceptability of such products is under investigation.[2]

References

1. "World-wide developments on fish flour," Nutrition Document R.8/Add. 10, WHO/FAO/UNICEF/PAG, August 1960 Meeting, Washington, D.C. Also A. Alcaraz-Bayan, and R. M. Leverton, "The Composition of Dilis Fish Flour," *Philippine Journal of Science,* 86:3 (1957) pp. 247-258.

2. Research work done on FPC at the Department of Fisheries Technology, UP College of Fisheries. Unpublished report provided by Rogelio O. Juliano, Dean, College of Fisheries, University of the Philippines, Diliman, Quezon City, Philippines.

POLAND

Experiments with preparation of FPC in Poland have been reported since 1964, but there is little evidence of the ways in which the product was used.

An early report describes the chemical properties of FPC made by treating fish fillets with dilute acetic acid and washing or by alkaline hydrolysis followed by precipitation with acetic acid. Both products were then dried and defatted with a mixture of ethanol and ethyl acetate.[1]

At the Swinoujscie Branch of the Marine Fisheries Institute (MIR) at Gdynia, a solvent extraction process was developed. This three-stage process relied on gasoline as the solvent for the first stage, ethyl alcohol "denatured with gasoline" for the second, and a water wash for the third.[2] The U.S. Bureau of Commercial Fisheries attempted to obtain more information on this process, but was unsuccessful. Another approach reported briefly in 1966 was an acid hydrolysis process to produce a deodorized product from fish meal made from salted eviscerated white fish.[3]

The degree of deodorization of FPC was tested in a series of experiments reported in 1968, with products made from herring stored on ice for varying lengths of time, then extracted with ethyl alcohol and ligroin (a nonpolar solvent mixture) in sequence.[4] (This investigation differed from a similar investigation by the BCF in that the former was concerned only with deodorization, as detected in biscuits made from the product, and the latter was concerned with all the properties that might depend on freshness of the raw material—yield, chemical composition, nutritive value, microbiological status, and organolyptic properties.)

Finally, a cooperative experiment was conducted by scientists from the Halifax Laboratory of the Fisheries Research Board of Canada and the Department of Fat Chemistry and Technology at Gdansk Technical University.[5] In this experiment the extraction efficiency of a three-stage, batch cross-current extraction of herring with isopropyl alcohol was investigated.

No substantive reports have appeared to indicate either what use was made of the materials produced in Poland or the current state of FPC activity in that country.

Recently, it has been learned that a large-scale FPC plant operates successfully in Gdynia, utilizing a process that was developed jointly with Unilever Company of Britain and Astra of Sweden. The material is said to have excellent functional properties and is widely used in processed meat products, such as sausages. The raw material is unutilized species of fish from the Baltic Sea. The industry is said to be economically viable and the FPC widely accepted.

References

1. Z. Podeszewski, "Changes in some N compounds during preparation and storage of fish flour," *Rocz. Nauk. Roln B.*, 84 (1964), pp. 685-710. (In Polish.)

2. Milan Dravanja, letter dated June 6, 1967, addressed to Dr. Norman P. Neureiter, Science Attaché, U.S. Embassy, Warsaw. Dr. Dravanja was Foreign Fisheries Specialist with the BCF Branch of Foreign Fisheries.

3. Anon., *Food Engineering,* November, 1966.

4. Z. Sikorski and H. Ziminska, "Effect of fish freshness of deodorization of fish protein concentrate," *Przemysl Spozywczy,* 22:9 (1968), pp. 405-406.

5. B. Drozdowski and R. G. Ackman, "Isopropyl alcohol extraction of oil and lipids in the production of fish protein concentrate," *Journal of the American Chemical Society,* 46:7 (1969), pp. 371-376.

PUERTO RICO

Dietary supplementation experiments with FPC have been performed in Puerto Rico, in an effort to increase the methionine content of typical rural diets that are deficient in this amino acid. In these experiments, three different products were used: dried, salted cod fish; deodorized VioBin FPC made from whole fish; and fish meal from Nova Scotia. In these experiments, the three materials performed comparably.[1]

Reference

1. J. A. Goyco and C. F. Asenjo, "La suplementación de la ración rural Puertoriqueña con próteina de pescado," *Arch. Latinoamer. Nutr.*, 17:3 (1967), pp. 241-251.

SAMOA (AMERICAN TRUST TERRITORY)

The use of FPC in the U.S. Trust territory of American Samoa would probably not have occurred had it not been for the appointment in 1969 of a new governor for the territory, John M. Haydon. At the time of his appointment Haydon had been serving as chairman of the oceanographic commission for the State of Washington; in this capacity he had been intimately involved with the efforts to win the competition for the siting of the FPC Experiment and Demonstration Plant (EDP) in Washington. (For details, see chapter 3.) In the course of his work with the Oceanographic Commission, Haydon also became associated with Congresswoman Julia Butler Hansen, in whose congressional district the EDP was eventually sited.

Upon his arrival in American Samoa, Governor Haydon learned that, while children in grades 1 to 12 received a daily hot school lunch, no nutrition supplementation was provided for the more than twenty-five hundred preschool youngsters (3½ to 5 years) on the island. It occurred to the governor that FPC, incorporated in some form of snack, might be ideally suited for use in a preschool feeding program, and he determined to make inquiries as to the possibilities of obtaining a quantity of the product for a pilot program. A major obstacle confronted by the governor, however, related to the continuing restriction of the FDA limiting the sale and/or distribution of FPC to one-pound packages. Without some sort of waiver on the FDA bulk packaging prohibition, a preschool feeding program using FPC would not be economically feasible.

With the influential assistance of Julia Butler Hansen, John Haydon was able to secure the cooperation of the National Marine Fisheries Service (NMFS) and a commitment by the Service for shipment of one ton of acceptable FPC. Armed with the NMFS promise, the governor traveled to Washington, D.C., in early 1972 and, on February 1 held a meeting with FDA officials. After describing the proposed nature and scope of the feeding program, the FDA agreed to waive the one-pound package limitation on the condition that the program take the form of a pilot study with the following results to be reported to the FDA: (1) documentation that the FPC used met FDA quality and safety standards; (2) narrative description of the logistical, processing, baking, and distribution phases of the program; (3) evaluation of the acceptability of the product to children; and (4) measurement of changes in heights and weights over time.[1]

After his return to American Samoa, Haydon convened a meeting of the relevant local officials from the island including the Director of Medical Services, Dr. Lowell M. Wiese, and the Director of Early Childhood Education (ECE), Dr. Betty Johnston. The group discussed the possibilities for using FPC in a preschool feeding program as well as arrangements for the visit of two experts from the NMFS College Park Laboratories, Drs. Bruce Stillings and Virginia Sidwell, the following July.[2] Pursuant to the meeting in the governor's office, both Johnston and Wiese submitted reports on March 9, 1972, commenting on the proposed program.

Dr. Johnston proposed that the Early Childhood Division serve a nutritious midmorning snack, in the form of FPC fortified cookies, to those children enrolled in the ECE program. She stated that the school lunch manager had assured her that he could bake cookies, containing 20 grams of FPC, in the present school lunch kitchen. This was predicated on the availability of adequate refrigeration and bug/rat-free storage capacity. She also noted that there were likely to be packaging, labeling, and distribution problems, but that these could be solved with adequate funding.[3]

Dr. Wiese's comments on the proposed program were a bit more circumspect. He stated that "analysis of presently available data does *not* support the hypothesis of any widespread protein nutritional deficiency in childhood or during the period of pregnancy."[4] He indicated that there had been no clinically valid cases of kwashiorkor or marasmus in children not otherwise ill or socially neglected or, for that matter, of other—often related—vitamin deficiencies such as beri-beri (thiamin) or eye inflammation, growth problems (riboflavin).[5] Wiese noted that although no totally adequate nutritional survey had been conducted on the island, a study by the Department of Medical Services of the heights and weights of three thousand Samoan children aged 0 to 19 revealed only slight differences from their U.S. counterparts. Thus, although he did not believe a protein deficiency problem existed, he felt there could be "no harm in giving the ECE children 20 additional grams of protein supplement daily" at least until adequate biochemical studies of the protein needs of American Samoa could be conducted. By Dr. Wiese's calculations, 20 grams of FPC daily for each ECE child would amount to a yearly requirement of approximately 12 tons.[6]

Progress on the project remained minimal from March until the arrival of the NMFS representatives in early July. The NMFS team had come to Samoa by way of Hawaii where they had stopped for consultations with officials at the University of Hawaii. While there, some doubt was raised about the need for such a project on American Samoa (especially as compared to recently independent Western Samoa) and as to the suitability of cookies as the vehicle for FPC in terms of the Samoan diet.[7]

On the basis of their work on the island, the NMFS team made a series of recommendations to the Government of Samoa relating to (1) equipment needs—mainly for scoops and scales; (2) personnel needs—three, full-time employees trained in the preparation and baking of the cookies; (3) kitchen facility problems—fumigation, screening, extermination and hygiene; and (4) distribution problems—testing of product lifespan in the tropical climate.[8]

In early August, Drs. Stillings and Sidwell sent a formal report on their findings which included suggestions for implementing a pilot study of a preschool midmorning snack program using FPC cookies. On August 24, 1972, a shipment of 950 pounds of FPC arrived by ship which, when added to a 50-pound shipment received earlier, brought the total FPC on hand to 1000 pounds.[9] Attempts were made during the fall to secure through the USDA Food and Nutrition Service the various pieces of equipment recommended by the NMFS team. Discussions were also held with the USDA agent about the possibility of subsidizing the midmorning snack program. These supplementary funding efforts were ultimately unsuccessful, however, because of USDA requirements that such a program must contain either hot soup or a beverage. Subsequent investigation by Samoan officials revealed that, due to the logistical cost of transporting milk or fruit juice to the island, the USDA subsidy, if obtained, would represent only a fraction of the cost of including these items in the snack.

By February 1973, the necessary kitchen equipment finally arrived and plans were initiated to test out the baking, packaging, and storing of a week's supply of FPC cookies in preparation for the beginning of the pilot study. On February 16, the School Lunch Supervisor and the Director of Early Childhood Education were present to check the first test batch. Several problems were encountered with insect infestation, sanitary procedures, and equipment.

A second test run on March 2 proved much more successful, and it led to the initiation of a full pilot study on March 12, 1973. FPC cookies were prepared and distributed five days a week for an eight-week period concluding on May 11, 1973. The ECE Teachers' Report provides some indication of the acceptability and success of the FPC nutrition intervention:

> The children were very happy and enjoyed eating the cookies. The school attendance has improved since the beginning of the Cookie Project. Three different kinds of cookies were served each day, but the children preferred vanilla and oatmeal. They looked forward to the time for the cookies to be served. At first, they wanted juice or milk, but later, they just settled for cookies. When the Cookie Pilot Study ended, the children were very disappointed. In fact, some of them have dropped out of school. We all feel strongly that this is a good program and it should be continued, if possible.[10]

The principal reason for the demise of the American Samoa FPC program appears to have related not to the acceptability of the cookies but to continuing programmatic difficulties in the United States. In the words of former Governor Haydon: "The kids liked the cookies, and the program would have been extremely popular if it had not been for the demise of the FPC plant at

Grays Harbor (the Aberdeen, Washington plant operated by the NMFS). When this took place, it was decided not to continue the program and build up a desire which could not be gratified."[11] No doubt, other factors also contributed to this decision, including the five reasons stated officially by the office of the governnor of American Samoa: (1) difficulty of distribution to certain centers because of remoteness and lack of regularly scheduled transportation; (2) lack of adequate storage facilities in all of the village centers; (3) pilfering of cookies or misuse of same by villagers; (4) lack of equipment to manufacture cookies in large volume; and (5) lack of adequate storage facilities for raw materials (that is, FPC) in large quantities.[12]

References

1. FDA internal memorandum, February 1, 1972.

2. Internal memorandum for the files, Department of Education, Early Childhood Division, Government of American Samoa, p. 1.

3. Memorandum of the Food Supplement Committee, Government of American Samoa, March 9, 1972.

4. Memorandum of the Department of Medical Services, Government of American Samoa, March 9, 1972.

5. *Ibid.*, appendix 1.

6. *Ibid.*, p. 1.

7. NMFS Internal Report, July 12, 1972, pp. 1-2.

8. *Ibid.*, pp. 7-9.

9. Internal memorandum for the files, Department of Education, Early Childhood Division, p. 2.

10. Report by the ECE Teachers on the FPC Pilot Study, Department of Education, Early Childhood Division.

11. Personal communication from John M. Haydon, former governor of American Samoa, August 22, 1975.

12. Personal communication from Earl B. Ruth, governor of American Samoa, August 5, 1975.

SENEGAL

In the middle to late 1950s, when FAO was concerned with testing FPC in various countries in Asia, Africa, and Latin America, Senegal was one of the countries in which feeding trials were conducted. The following excerpt from a PAG document summarizes the situation at the time.[1]

Sénécal and co-workers recently described[2] some preliminary experiments in Dakar on about thirty infants and young children (ranging from 5 months to 9 years of age, but most of them below 2 years) with a few non-deodorized fish flours locally produced on a small experimental scale, after some toxicity tests . . . had been carried out in the Institut Pasteur apparently with negative results.

The fish flours were incorporated in a number of local diets; the fish flours were tolerated well. Because of the limited quantities of fish flour available it was not possible to do long-term experiments in the hospital (one and one-half months). Only two children (2 years of age) received fish flour during their whole stay . . . It is stated that these fish flours are suitable as supplementary foods to infants from 6 months on.

From these comments is appears that what was being fed to the children was fish meal.

No other information is available on FPC activities in Senegal.

References

1. Nutrition Document R.8/Add. 1., WHO/FAO/UNICEF/PAG, March 15, 1956.

2. Sénécal *et al.*, "Bull. et Mémoires de l'Ecole prep. de Méd. et de Pharm. de Dakar. Part II," 1954.

SOUTH AFRICA

South Africa has the longest history, in modern times, of attempts to prepare and use fish protein concentrate as a dietary supplement. Indeed, the idea of producing a deodorized fish meal suitable for human consumption probably originated with Dr. G. M. Dreosti in 1937. Dr. Dreosti, (at that time, Officer in Charge of Dehydration and Cold Storage, Low Temperature Laboratory, Capetown), began investigations aimed at producing a tasteless and odorless fish flour in collaboration with F. S. Mendel, of the firm Concentra Ltd. Their aim was to use this fish flour to enrich cereal products intended for human consumption. By 1939, they had succeeded in producing a bland product by extraction of white fish meal with ethyl alcohol. This product, however, tended to regain a slight fishy odor and flavor after prolonged storage. Because of the nature of the process the product was expensive and there was not much public interest in it.

With the intervention of the war the project was discontinued until 1951 when work was resumed after the Minister of Health announced plans to enrich staple foods with protein supplements and became very interested in the FPC experiments. By this time the Fishery Industry Research Institute (FIRI) had been established with Dr. Dreosti as director. FIRI then began an extensive investigation of the production of FPC not only from white fish meal but also from the less expensive, brown-fleshed oily fishes, pilchard and maasbanker.

About the same time Marine Oil Refiners of Africa, Ltd. (MOR) also began to work on the extraction of pilchard meal, with the product intended for use in brown bread. By September 1951, both FIRI and MOR felt they were able to produce a satisfactory product, and early in 1952 MOR set up a pilot plant to produce a fish flour made by extraction with ethyl alcohol under pressure at temperatures of 120–150° C. FIRI, meanwhile, used an ethanol extraction at 80–85° C. The MOR product was used by FAO in acceptability tests in Chile in 1953 with positive results.

It was established subsequently, however, that the high-temperature, high-pressure process resulted in a product with impaired nutritive value. Also, because of problems with solvent recovery, the MOR product retained undesirable odors and flavors. MOR abandoned its project and the pilot plant was then acquired by the South African Fish Meal Producers Association which used the plant to produce FPC by the FIRI process beginning in 1956. By the end of 1958, more than 1000 tons had been produced. The product was used to enrich brown bread (2 percent FPC on a dry basis) in a project sponsored by the Department of Nutrition. The effectiveness of the bread enrichment scheme was the subject of a detailed study, and it culminated in the publication of the book, *Food Enrichment in South Africa,* by the National Nutrition Research Institute (NNRI) of the CSIR.[1]

The review of the bread enrichment scheme indicated that "The only section of the South African population in which there is evidence of a deficiency primarily of protein is the group comprising Non-European infants and young children aged from about 6 months to 5 years . . . there is no clear-cut evidence in the case of older age groups indicating the widespread incidence of a specific syndrome in which protein deficiency has been shown to be a primary importance."

In view of this, and the feeling that the target group of infants and young children consumed significant amounts of bread, the enrichment scheme was terminated. It was replaced by distribution of dried skimmed milk (DSM) through local clinics. The FPC pilot plant was shut down and the fishing industry lost interest in producing FPC.

While they lasted, the supplies of FPC were used in "Pro-Nutro," a high-protein food developed and manufactured in South Africa.

The NNRI subsequently developed a protein-vitamin-mineral food supplement (PVM) that contains FPC.[2] This supplement is currently being produced with imported FPC by two firms in Pretoria. The Department of Health is using PVM as a substitute for DSM on an experimental basis in the feeding scheme referred to. No decision has been made about the eventual outcome of this scheme or about the future use of PVM.

References

1. Much of the material in this section was drawn from three sources:

G. M. Dreosti and R. P. van der Merwe, "The manufacture of fish flour in South Africa," Fishing Industry Research Institute Progress Report No. 64, in *The South African Shipping News and Fishing Industry Review,* August 1961.

J. P. De Wit, personal communication, 1975, Dr. De Wit, Director of the National Food Research Institute, Council for Scientific and Industrial Research at Pretoria, very kindly supplied a host of materials on the history of FPC in South Africa, and an excellent review of that history in a letter dated September 2, 1975. The authors wish to take this opportunity to acknowledge with thanks his cooperation.

Food Enrichment in South Africa, National Nutrition Research Institute, South African Council for Scientific and Industrial Research, Pretoria, 1959.

2. For Information on PVM and its uses, the following references should be consulted:

J. J. Dreyer, D. B. du Bruyn, C. P. Van Wyck, and T. R. J. Concannon, "Background information on the development at the NNRI of a supplement for the predominantly maize diet," *South Africa Medical Journal,* 39 (1965), pp. 1167–1169.

J. P. De Wit, "Methods of combatting malnutrition," Paper read at the Symposium on Proteins and Food Supply Bloemfontein, April 1968.

P. F. Pretorius, "Clinical Nature and extent of protein malnutrition in South Africa," *South Africa Medical Journal,* 42 (1968), pp. 956–958.

J. J. Dreyer, and D. B. du Bruyn, "Composition of a food mixture developed at the NNRI for Supplementation of predominantly cereal diets," *South Africa Med. Jour.,* pp. 600–604.

J. J. Dreyer, "Supplementary effect of the NNRI food mixture on cereal proteins," *South Africa Med. Jour.,* pp. 604–610.

D. B. du Bruyn, "Mineral and vitamin efficiency of the NNRI food mixture," *South Africa Med. Jour.,* pp. 611–616.

W. Wittmann, J. G. Prinsloo, and H. Kruger, "Efficiency of the NNRI food mixture as tested on kwashiorkor patients," *South Africa Med. Jour.,* pp. 959–961.

F. Schweigart, "The preparation and keeping quality of the NNRI food mixture (PVM)," *South Africa Medical Journal,* 43 (1969), pp. 328–331.

V. A. Cachia, "Acceptability and usage of the NNRI food mixture (PVM)," *S. Africa Med. Jour.,* pp. 336–338.

D. B. du Bruyn, "The cause of the myocardial 'hyaline degeneration' induced in rats fed predominantly maize diets supplemented with either the NNRI food formulae or skimmed-milk powder," *S. Africa Med. Jour.*, pp. 341-344.

3. A representative group of papers on the South African FPC production follows:

Aneri, "Flour from maasbanker," *Fish Newsletter* (Australia), 15 (1956), p. 19.

G. M. Dreosti and R. P. van der Merwe, "Fish meal. 18. Fish flour," Annual Report, Fishing Industry Research Institute, Cape Town, South Africa, 9 (1955-1956), pp. 32-33.

G. M. Dreosti and R. P. van der Merwe, "Fish flour. 2. Factors influencing rate of extraction," Annual Report, Fishing Industry Research Institute, Cape Town, South Africa, 10 (1957), pp. 34-36.

G. M. Dreosti and R. P. van der Merwe, "Fish flour. 3. Hold-up fat," *Annual Report,* FIRI, pp. 36-37.

S. G. Wiechers, F. Schweigart, and M. K. Rowan, "Microbiological process for the production of a bland fish meal," *CSIR Res. Rep.*, No. 179, 1960.

G. M. Dreosti, R. P. van der Merwe, and J. J. Dreyer, "Fish flour—Technological Developments in South Africa," FAO International Conference on Fish in Nutrition, September 19-27, 1961, Washington, D.C., in *Fish in Nutrition,* (London: Fishing News [Books] Ltd.), pp. 425-431.

4. P. J. Pretorius, and A. S. Wehmeyer, "An assessment of nutritive value of fish flour in the treatment of convalescent kwashiorkor patients," *American Journal Chemical Nutrition,* 14 (1964), pp. 147-155.

SPAIN

The only indication of interest in FPC for the eventual feeding of humans is found in three papers reporting results of animal feeding tests.[1,2,3]

References

1. J. Gallego, F. Morales, E. Ronda Lain, and L. Ruiz Perez, "Supplementation of d-l-methionine and of l-lysine using a corn-soy-bean diet with and without fish powder and powdered milk in young chickens," *Rev. Patronato Biol. Anim.*, 7:3 (1963), pp. 61-68.

2. J. Bello, C. Rodriguez, and J. Larralde, "On the nutritional indices of proteins in Scomber scombrus, Thunnus thynnus," Germoalalunga and Euthynnus Pelamis. An. Bromatol., 8:4 (1966), pp. 385-392. Feeding tests in which young rats are fed with fish flour made from the above fish species.

3. J. Bello, J. Larralde, and C. Rodriguez, "On the biological value and other nutritional indices of Mustelus mustelus and Raja clavata proteins," *An. Bromatol.,* 16:4 (1964), pp. 475–483. Rat feeding tests with FPC made from the above fish species.

SRI LANKA (FORMERLY CEYLON)

We have one account of a successful study in which two groups of convalescent children were given a diet containing FPC.[1]

Reference

1. T. W. Nikramanayake, C. De. S. Nijesundra, C. C. De Silva, and N. N. De Silva, "Supplementation with fish flour of a rice diet fed to convalescent children," *Journal of Tropical Pediatrics,* 12 (1966), pp. 7–14.

SUDAN

In 1972, an extensive survey was conducted in Southern Sudan to determine the suitability of FPC as a protein supplement in the relief and resettlement operation then under way.[1] The survey was made possible by the cooperation of the National Nutrition Council FAO Committee of Norway; FPC was donated by the Swedish and Norwegian Red Cross. Fourteen tons of Type A and Type B FPC were supplied for this survey, and a total of approximately 6150 people tested foods supplemented with it.

The survey report noted that, because of food patterns in Southern Sudan, the bland, odorless, tasteless product (Type A) was not suited for dietary supplementation. The less refined product (Type B), however, was accepted well when mixed in the broth commonly consumed during meals.

In view of the fact that there was no sign of impending famine, or even of a serious lack of a variety of foods sufficient to support a balanced diet, no further move was made to import FPC.

Reference

1. Cato Aall, Report on acceptability and possible use of fish protein concentrate in the Southern Sudan, August 26, 1972. (Copies of the report can be obtained from: Ministry of Health, P. O. 88, Juba, Southern Sudan, or, National Nutrition Council/FAO Committee, P.O. Box 8139, Oslo Dep, Oslo 1, Norway.)

TANZANIA

FPC activity in Tanzania began in the 1960s in an effort to supplement the starch-rich diet of most Tanzanians and to achieve maximum utilization of available fish stocks. This course was taken by the government, instead of encouraging the consumption of fish, because of "taboos and prejudices towards certain types of fish" and the feeling that an enriched corn-flour meal could be marketed successfully if no specific indication of the source of the enrichment were made.[1]

Experiments were made with a solvent extracted product, with the assistance of Dutch technicians. The investigation is still continuing, with as yet no satisfactory results.[2]

References

1. John J. Kambina, letter to W. M. Chapman dated September 5, 1967. Mr. Kambina was Director of Fisheries, Dar es Salaam.

2. A. C. Mosha, Director, Food Science and Technology Department, Tanzania Food and Nutrition Centre, personal communication, 1975.

UGANDA

In a bulletin of the FAO/WHO/UNICEF High-Protein Food Program, it is reported that feeding trials were conducted in Uganda in 1957.[1] Dr. R. F. A. Dean, Director, Medical Research Council, Group for Research in Infantile Malnutrition, at the Mulago Hospital in Kampala supplemented the diets of infants recovering from kwashiorkor, and those of preschool and school-age children. Using biscuits prepared from groundnuts, wheat flour, maize flour, edible oil and sugar, with and without supplements of dried skimmed milk or fish flour, he tested the nutritive value of the foods, and studied the growth of the children.

There is no record of current interest in FPC in Uganda.

Reference

1. FAO/WHO/UNICEF High-Protein Food Program, News Bulletin No. 1., October 1957.

UNITED KINGDOM

Research and/or development efforts on FPC in the United Kingdom seem to have been confined exclusively to minor efforts in the private sector. The reason for this was discussed at the June 1961 meeting of the PAG. Reference was made to a UNICEF meeting with the representatives of the Torry Research Station (Aberdeen, Scotland) of the Ministry of Agriculture, Fisheries and Food, convened to discuss "the most suitable locations for the production of fish flour." UNICEF noted that "the opinion was expressed by others in the UK that the production of fish meal in the UK is no more than an economic means of dealing with fish that is either unfit for human consumption or will be by the time it has passed through the usual channels of distribution." The UNICEF report goes on to say, "They believe that there is not a supply of white fish available in the U.K. for the production of fish flour of the quality desired by the U.N. agencies," and concludes that "there is an expressed reluctance on the part of U.K. authorities to divert research monies to a project on the development of a fish flour for human consumption that appears to have little advantage to British industry."[1]

This official attitude has remained unchanged, although some government-sponsored work on a hydrolyzed fish product is apparently underway currently.[2] Another minor exception was the work of the Herring Oil Research Station (Department of Scientific and Industrial Research) at Yarmouth. This organization developed a method of simultaneously dehydrating and defatting herring as means of offering the fishing industry an alternative to fish meal processing.[3] The final outcome of this research is unclear, although in view of the stand of the Ministry of Agriculture, Fisheries and Food noted above,[4] it is likely that the work has long since ceased.

Aside from this, the only other activity that can be identified has been in the private sector.[5] Rose, Downs and Thompson Ltd. of Hull, for example, was reported to have done "extensive research on the development of odorless and tasteless fish flour," presumably in the 1950s. The work was reported to be "quite successful but was discontinued when they were advised there was a lack of a market." A sample of their product, prepared from white fish in 1958, was submitted to UNICEF in 1960 and was judged to have an acceptable flavor and odor.[6]

Interesting discussions of the British attitude toward the FPC concept and outlook are given by J. A. Lovern of the Torry Research Station.[7,8]

References

1. Nutrition Document R. 8/Add. 11, WHO/FAO/UNICEF/PAG, June 1961 Meeting, New York, New York, p. 4.

2. Ian M. Mackie, personal communication, 1975, Mr. Mackie is Director, Torry Research Station, Ministry of Agriculture, Fisheries and Food. In a letter dated September 2, 1975, Mr. Mackie states, "There has been no FPC research or development in the UK. As far as we know the Caledonian Fish

Meal Company produced a high grade fish meal by a conventional fish meal process but only on a pilot scale." The purpose of this activity was "to demonstrate feasibility of using the fish meal process to produce a dried protein suitable for human consumption. It was not marketed."

3. "Fish Protein Concentrate (FPC) Program Review and Plan, Part I," Report prepared for National Marine Fisheries Service, NOAA, Washington, D.C. by Northrup Services, Inc., p. G–10.

4. See note 2.

5. For example:
a. British Patent No. 727072, 1965, "Powdered fish production." This patent describes a cross-current solvent extraction of fish with acetone and ethyl alcohol.
b. British Patent No. 1, 108, and 188, 1966. Purification and concentration of fish protein. This patent describes an alkaline hydrolysis procedure for separating and concentrating the proteins.
Other British patents have been issued, but they concern processes developed in other countries.

6. Nutrition Document R. 8/Add. 10, WHO/FAO/UNICEF/PAG, August 1960 Meeting, Washington, D.C., p. 3.

7. J. A. Lovern, "Problems in the development of fish protein concentrates," *Proc. Nutr. Soc.*, 28:1 (1969), pp. 81–85.

8. J. A. Lovern, "General outlook for edible fish protein concentrate," *Advan. Chem. Ser.*, 57 (1966), pp. 37–41.

URUGUAY

The most significant FPC-related activity in Uruguay took place at the Universidad Republicana. Using a proteolytic yeast isolated from the liver of a marine fish, Professor Victor H. Bertullo devised a hydrolysis process for producing a soluble FPC. A patent was eventually granted on Bertullo's process[1] but, as far as is known, no commercial manufacture has ever been attempted.

During the period in the BCF FPC program prior to the submission of the Food Additive Petition to the Food and Drug Administration, Professor Bertullo visited the College Park Laboratory to supervise the preparation of FPC by his method. The results, however, were inconclusive, and the BCF never followed it up.

Reference

1. Victor H. Bertullo, U.S. Patent 3,000789, September 19, 1961.

USSR

Interest in FPC in the USSR stems from the early 1960s. The earliest report available (1954) describes a process involving alkaline hydrolysis followed by acetic acid precipitation and extraction. Lean fish were extracted with ethanol, and fatty fish with ethylene dichloride.[1]

In a later publication a method utilizing heat and pressure is mentioned in connection with the manufacture of fish flour (fish meal?),[2] and in 1973 an article appears discussing the amino acid composition of FPC from Caspian sprat.[3]

References

1. A. I. Yuditskay, "Dry protein from low-grade small fish," *Rybnoe Khoz.*, No. 9 (1954), pp. 56–59.

2. V. I. Treshcheva, "Changes in the vitamin B, B_2 and nicotinic acid content in the preparation and storage of fish flour," *Izvest. Vysshikh Uchebn. Zavedenii, Pishchevaya Tekhnol.*, 1 (1963), 62–65.

3. Y. A. Chermin, "Amino acid composition of the fish protein concentrate from caspian sprat," *Vopr. Pitan*, 31:5 (1973), pp. 45–48.

VENEZUELA

In 1961 (and earlier), a series of experiments were performed on the nutritive value of FPC compared with fish meal.[1] Solvent extracted FPC, obtained from the United States, and whole sardine fish meal, made in Venezuela, were added to wheat flour and bread in a rat-feeding study. The one unique result reported is that FPC gave better growth in the test animals when added to the bread after baking rather than to the flour before the bread was baked.

Reference

1. W. G. Jaffe, "New studies of the nutritive value of fish flour and its supplementary effect on wheat flour and bread," *Arch. Venezol. Nutr.*, 11 (1961), pp. 191–204.

PART IV
AN ANALYSIS OF THE INTERNATIONAL FPC EXPERIENCE

Chapter 11
Policy Implications of the FPC Experience

When the FPC experience of the past twenty-five years is viewed as a whole, certain aspects are particularly significant in terms of both food and nutrition policy formulation and the development of new foods or food components. It is significant that most of the positive results of the FPC story concern the fish resource and the processing technology and are beneficial chiefly in the context of the economies of developed countries and in the context of long-range development. The negative aspects, on the other hand, are primarily related to the complex dimensions of the short-term, urgent malnutrition problem in low-income countries and to the failure of FPC, both as concept and product, to be effective within this framework. While the positive effects are not insignificant, the negative aspects demand the attention of all those policymakers, planners, and technologists whose interest lies in improving the nutritional status of impoverished populations.

POSITIVE ASPECTS OF FPC DEVELOPMENT

Extension of Food Resource Base

Increasing population, industrialization, and urbanization have, of necessity, resulted in the development of mass food production and distribution technologies. Mass production and distribution of food, in turn, have reduced the variety and number of plant and animal species used in food manufacture. Goals of higher yield, disease resistance, better distribution, ease of husbanding, harvesting, and processing, as well as higher nutritive value per acre, have all favored certain species for utilization rather than others. Today only a fraction of the raw materials available for the manufacture of food is utilized and acceptable as a food source. (While this is particularly true for industrialized nations, production goals like those mentioned are gradually being transferred to the food systems of developing countries as well.) Of some two million known species of animal only about fifty have been domesticated and contribute to human food. Of the many thousand existing species of edible fish only one or two dozen constitute the major amount of fish consumed today worldwide.

Enough data exists on the present state of world fisheries to convince us that almost all the major traditionally preferable species have been overfished practically everywhere in the world. Data also exist, however, that point to very substantial stocks of food-grade schooling fish which, though harvestable by existing methods, remain either unutilized or underutilized, principally as a result of market patterns in the industrialized world and rigidity of cultural preferences.

The development of FPC, which can be produced from any edible fish species, has served to indicate how this resource base of raw material can be expanded for the manufacture of food long before this need has actually become topical. The resource can indeed be further extended now by drawing upon such species as krill, squid, and capelin, to mention only a few, that have hitherto not been part of the world's larder.

Wider Utilization of Fish in Nontraditional Forms

Fish in western countries has been traded and consumed in modern times primarily in its original form, the form in which it is recognizable as fish. This has had the effect of focusing fish harvesting efforts on species for which a known market demand exists and of reinforcing the reluctance of food manufacturers to develop food processing methods designed to utilize fish in new or alternative forms as a food component similar to processed milk, egg, or wheat.

In the process of resolving the widespread debate, particularly in the United States and Canada, over the more controversial aspects of FPC development—the so-called "filth" issue, the question of acceptable species, and the controversy over packaging and labeling—wider possibilities for the use of fish as food, especially in new and/or previously unutilized forms, became known to and accepted by food manufacturers and the public. Today, the popularity of so-called "fish burgers," for example, is widespread in the quick service food market. And even the U.S. FDA now recommends the use of FPC as a means of upgrading the protein value of certain vegetable-based meat analogs.

Thus, the development of FPC has contributed to a change in public attitude toward fish as a food component, helping to promote the idea that fish can be consumed as an invisible ingredient, that is, in combination with other food elements rather than exclusively in its whole form. The net effect of these changes in attitude is an increased inclination on the part of the food industry to utilize available fishery resources more effectively.

Relaxation of Regulatory Attitudes Toward Novel Food Sources

We have suggested that one piece of tangible evidence supporting the wider use of fishery resources in the United States is the endorsement by the FDA of fish protein concentrate as a food fortificant. In addition, FDA staff members now recognize formally the usefulness of products like FPC in view of the restricted supply of traditional sources of protein.

In part due to the advances in food-processing technology (of which FPC was an important manifestation) and in part to the skyrocketing prices of many of the more traditional sources of protein, the recent trend in food regulatory policies in the United States and other industrialized nations has generally been toward a more liberal interpretation of what constitutes a human-grade foodstuff. This has occurred despite parallel improvements in the technology for detecting the presence of minute quantities of foreign matter (filth) in food. The trend is indicative of a slowly growing recognition by the regulatory bodies in the developed world that (1) it is not the responsibility of government to judge food on aesthetic and social grounds since the changing economics of food supply may cause periodic redefinitions of what is considered acceptable; (2) increased interaction between cultures, as a result of the sophistication of modern transportation and communication systems, will bring into use foods that were not previously an accepted part of a society's food consumption patterns; and (3) continuing advances in the state of the art of food science will make available new products, for which there is no legal or administrative precedent, whose suitability for human consumption must consequently be judged largely on the basis of functional properties without indicating concerns about wholesomeness.

While these tendencies toward change in regulatory policy are both significant and laudable within the context of the developed countries, they have a direct impact on the developing world as well. Many developing nations model their food regulatory statutes directly on those of the United States or other developed countries because they have neither the technical capability nor the funds to conduct the complex analyses necessary to establish their own statutes. Nationalistic pride is also a factor that may act to prevent the use of a product in one country—no matter how appropriate the product's design—simply because it has not been approved for use in its country of origin. The relaxation of regulatory attitudes may thus have major impacts, in both developing and developed nations.

Expanded Production of Protein Concentrates

The development of FPC was important in that it has made available a stable, easily digested high-grade protein concentrate that can be used effectively in the clinical treatment of acute protein deficiency diseases. Its acceptability as a protein-boosting ingredient in food for general use has been proved in the case of Astra's FPC, which has been incorporated in small quantities in bread sold on the retail market in Sweden and distributed by the World Food Program in food-for-work projects. The U.S. FDA has recommended its use in the production of meat analogs from vegetable sources for the purpose of raising the protein value of these engineered foods to the level of the foods they imitate. In short, all the research and development efforts put into FPC have resulted in a product that works well within certain contexts. The development of a protein concentrate made from fish was a precursor and a reference point for other efforts to concentrate proteins from unusual sources. Single cell and leaf proteins, like FPC, are achievements of sophisticated technology whose range of application has so far been limited but whose effectiveness in specialized situations is proved.

Development of New Food Processing Technologies

The variety of technical problems and specifications confronted in the course of research and development for the manufacture of whole fish protein concentrate has resulted in the establishment of a new branch of food processing and quality control technologies. Among the more notable and potentially useful of these developments are the following: (1) methods for the storage and preparation of comminuted fish; (2) techniques for the separation of fish muscle; (3) technologies for solvent extraction; (4) processes for desolventization; (5) means of solvent deodorization and recovery; and (6) techniques for the isolation of protein by enzymatic and other methods.

These spin-offs and generational improvements provide substantive evidence that many of the technical problems encountered—and, in many cases, resolved—in the effort to develop satisfactory industrial-scale manufacturing procedures have engendered profitable applications beyond the production of FPC. In fact, it now seems likely that these technological advances, spawned by FPC programs, may have a far greater impact on world eating habits and absolute food supply than the FPC development efforts themselves.

Demonstration Effect of FPC Development

Another closely related, positive consequence of the FPC development effort has been its demonstration effect, particularly for government bureaucracy. The myriad of delays, cost miscalculations, and technical pitfalls that were encountered in the course of product gestation has provided yet another illustration of the difficulties inherent in moving a product or process from the realm of the technically possible to the status of cost-effective production.

The short-term consequence of these programmatic realities was a dampening of both public and private sector interest in pursuing other projects of a similar nature. However, the long-term result may be identified to some degree with the growing recognition within government (at least in the United States) of the need for and value of comprehensive planning and cost-benefit analysis (such as the introduction during the mid-1960s of Planning Programming Budgeting Systems within U.S. government agencies) before proceeding into areas of great technological uncertainty. Of course, we do not mean to suggest that the FPC experience in and of itself has been responsible for such far-reaching managerial innovations. But the cumulative effect of the lessons learned from the U.S. FPC program may have contributed to a more circumspect, rational attitude toward future ventures.

Precedent for the Allocation of R&D Resources to Problems in the Developing World

One of the rationales most frequently offered in support of the commitment of public funds to the development of FPC technology was its potential usefulness in the alleviation of protein malnutrition in the developing world. While this goal was never proposed as the *sole* justification for the program, even its mention was significant in that it represented a new direction for R&D policy in the developed world. It is well known that countries such as the United States, Canada, and Sweden have for many years been transferring technology originally developed for the benefit of the developed country to the LDCs, generally in the form of technical assistance and/or capital loans and grants. But the case of FPC is perhaps one of the first documented, postwar examples of a technology that was created ostensibly for the benefit of the developing countries, yet was supported chiefly through public R&D funds in the developed world.

In reality, of course, the FPC programs were not primarily created for the benefit of the developing countries but rather to advance the political and economic interests of certain elements of the private and public sectors in the industrialized states. In the case of Sweden (which should, strictly speaking, be eliminated from consideration in this regard since no public funds were used there directly in the course of development), this conflict between stated and unstated policy was avoided through private development by the Astra Group for the product's likely future profit potential. Yet even in this example the private development program was sustained in part by major purchases of FPC by the Swedish government for use in the World Food Program. In any event, in the United States and Canada at least, federal funding for FPC represented a precedent-setting departure from prior foreign aid policies that had emphasized the utilization of technology already extant in the donor country.

In the period since the decision to support the development of FPC in the United States and elsewhere with public funds, there have been a number of additional public sector programs created to assist directly in the problems of development. Such efforts have been undertaken not only in the area of food science (for example, corn-soy milk, wheat-soy blend, amino acid fortification) but also with regard to the development and transfer of so-called "appropriate technologies" in fields such as agriculture (farm implements and techniques), water resources (tubewells), and manufacturing (cottage-type industry). Most of these initiatives have been more successful than the FPC venture, however, in that they were designed with consideration for the kinds of social, cultural, and economic variables that the FPC developers generally ignored.

Qualified Success of Astra

The success of Astra Nutrition in producing FPC for both human and animal consumption demonstrates that, given adequate inputs of time, money, technical expertise, and commitment, technological problems can be overcome and new foods or food ingredients can be developed from previously un- or underutilized sources. Two elements of the Astra endeavor are most significant in their implications for future development of products like FPC. First, Astra's experience as a leading manufacturer of pharmaceuticals meant that it was familiar with the kind of long lead time and planning necessary for the research, development, and testing of a previously unknown product. Such time requirements did not represent a limiting constraint for Astra as they did for most of the other FPC developers. Second, the fact that the Astra Group (of which Astra Nutrition is a small division) is a wealthy and diversified corporation that could readily provide the financial support crucial for this long-term, high-risk endeavor and easily absorb the expense and technological uncertainty of research and development of FPC.

Astra's success is one of development and production, and in these areas it is exemplary, as is the favorable disposition of the Swedish government and its regulatory agencies toward the endeavor. All of these factors have not succeeded, however, in creating the markets for FPC that Astra had originally intended. Although the majority of Astra's food-grade FPC is used by the World Food Program, the purchase is made by the Swedish government for donation and the quantities of FPC are quite small in comparison with other commodities. Astra has recently curtailed production of food-grade FPC in order to dispose of accumulated stocks that are partly the result of a slowdown in UN use. Given the time and cost constraints of the Astra model of development, it is likely that the production of FPC and of products like it will remain the domain of wealthy companies in industrialized nations, as will the market for them.

NEGATIVE ASPECTS OF THE FPC EXPERIENCE

Malnutrition Ill Defined

During the major period of FPC development, malnutrition was considered primarily as a protein deficiency disease, and the notion that the "protein gap" between well- and malnourished nations could be bridged by developments in food technology had great appeal. This attitude offered intellectual and moral support to developers of widely varying motivations in their search for funding and interest for FPC work. Intellectually, the prevalent scientific "theory" underlined the need for such a product, and morally the humanitarian value of solving the problem of malnutrition was undeniable.

The popularity that the protein gap idea enjoyed for so many years, however, was based largely on its simplistic view of a complex situation. If protein deficiency was the problem protein was the obvious and quite attainable solution. Thus, the idea of utilizing a product like FPC as a solution to malnutrition was as simplistic as the protein gap concept itself. The vast resources of the sea, developers said, could be tapped for the production of huge quantities of FPC. Distinctions between protein ingredient and food or between production and actual consumption of a product were rarely made in the hyperbole surrounding FPC development.

The appeal of the FPC solution clearly lay in the fact that it placed few demands on any sector other than the technological sector during an era when faith in technology was almost unshakable. In most cases that we have considered, FPC development was appealing because it offered the hope that malnutrition could somehow be alleviated without changing the social, economic, and political balance which more often than not constituted the supporting framework for its existence. This has been termed by many the so-called quick and painless "technological fix." Though the definition of malnutrition has broadened to encompass not only protein deficiency but also calorie deficiency and aspects of economic and social causation, the appeal of technological "solutions" is still strong, particularly for those who reject the idea of significant social and political change as a means to more equitable distribution of income and, consequently, of food.

Design of FPC as a Curative Agent

As might be expected, "technological fixes" based on misinformed judgments concerning the nature of the malnutrition problem yield inapplicable and/or ineffective solutions. Fundamental to an understanding of why FPC was an inappropriate solution to the problem of protein malnutrition is an understanding of the model adhered to by the various developers. Malnutrition was viewed at that time in terms of the so-called medical model and its *solution* was also sought in the same paradigmatic terms: Develop a curative agent that would be applicable wherever the "disease" of protein malnutrition occurred.

It is now apparent, however, that there are profound differences between a product such as FPC and a typical curative agent. For one thing, even the very name fish protein concentrate meant that the cloak of anonymity that surrounds the esoteric contents of most medicines did not apply to FPC. This problem might well have been overcome, however, had not a more fundamental obstacle been overlooked in the course of FPC development. Although significant progress had been achieved in developing technological approaches to particular *micro*nutrient deficiencies (such as the vitamin A fortification of milk and the iodine fortification of salt), the attempt to devise a technological solution to a *macro*nutrient deficiency such as protein (even though the calorie problem had been overlooked almost entirely) entailed problems of a completely different magnitude. Whereas micronutrient deficiencies require only *micro*dosages and therefore lend themselves to a "medical-type" intervention design, macronutrient deficits (whether protein or protein-calorie) require consumption in relatively large daily quantities (an ounce or more). As a result, "medical-type" intervention as a response to macro deficits is entirely inappropriate. Protein or protein-calorie interventions can, in fact, be taken only for what they actually are, namely, sources of food, with all of the social, cultural, and economic preferences and constraints that this involves.

In addition, there were a number of other significant discrepancies between the use of a typical curative agent and the use of FPC: (1) whereas medicines are required only for the duration of a disease, the protein concentrate must be consumed for as long as the protein deficiency exists. If the problem is long-term, *chronic* protein malnutrition, where large portions of the population are suffering from first or second degree deficiency (rather than *acute*, third degree malnutrition precipitated by famine or other disasters), it would be necessary for such individuals to continue to consume the protein for an indeterminately long period of time, that is, until the social, economic, political, and environmental factors responsible for the lack of adequate, nutritious food supplies had been modified; (2) whereas medicines can be administered either in a central location or taken by the patient, often in premeasured dosages, protein concentrates must be mixed either into a staple food item that reaches all of those in need or, more likely, mixed into foods in an unregulated fashion by consumers themselves; and (3) whereas many people may accept the positive benefit of the components of drugs on faith, a protein concentrate such as FPC has distinct problems of consumer acceptability due to wide public knowledge concerning the product's raw material components.

Thus FPC, as a response to the putative problem of protein deficiency, was unable by its very design to have the universal application of curative agent and, therefore, the use of the medical model for developing the product was clearly inappropriate. This fact, however, did not deter promoters from attempting to develop FPC as a medicine anyway. Many of the crucial functional properties of the resulting product were thus determined *as a consequence* of the decision to produce a curative agent. For example, the decision that FPC was as far as possible to consist of nothing but a high-grade native protein and was to have organoleptic properties suitable for combination with a wide variety of foods without changing their appearance, odor, or flavor occurred as a de facto result of adherence to the medical model.

Such rigorous characteristics required the development of complicated, capital-intensive, large-scale processing technologies (including chemicals, equipment, and control mechanisms), the training of sophisticated and highly paid labor, and the elimination of a variety of serious bureaucratic, political, and economic obstacles. These requirements drove the cost of the final product to unanticipated—and ultimately unacceptable—heights. The "curative agent" approach demonstrated a significant failure by the developers to grasp the complexity of designing a product for malnourished populations with widely varying needs and in widely varying locales.

Conflicting Motivations for FPC Development

In retrospect, the causes for the failure of FPC programs are so clear that it is difficult to comprehend how any well-informed individual, much less an organization or institution, could not have seen the implications of at least *some* of the problems cited here. Even allowing for the advantage of hindsight, the fact is that many of the problems were *not* all that opaque; they were simply misinterpreted (consciously or naively) by the individuals or organizations involved. Such misinterpretation often arose from the fact that much of the motivation for FPC development had little or nothing to do with the ostensible and well-publicized goal of using FPC in the alleviation of protein malnutrition. While this humanitarian goal proved useful in efforts to gain time, funds, and political approval for development efforts, it often served as a convenient mask for the varied objectives of the different development groups.

FPC did not, of course, simply appear on the scene fully developed. Like the vast majority of new technological concepts, it evolved, incrementally and unevenly, in many places and for a variety of reasons. The outcome of the worldwide FPC programs reflected the convergence of a complex array of entrepreneurial agents, each acting from their particular agenda of motivations and objectives. Not surprisingly, the mature character of FPC technology tended to reflect these pluralistic origins.

We are not using the word entrepreneur to mean simply the traditional, pioneering capitalist (although the story contains its fair share of those) but rather to refer to the variety of individuals and groups who exercised strong and historically significant influence on the direction, rate, and scope of the product's development. In this sense we distinguish five principal entrepreneurial groups active in the FPC story: industrialists, politicians, bureaucrats, scientists, and technologists.

True to the best entrepreneurial tradition, and in a manner quite appropriate to free market economies, *industrialists* who were enthusiastic about FPC perceived it as a potential source of profit. Consequently, they argued strongly for a minimum of government involvement and regulation in order to produce FPC quickly and with the fewest possible cost-increasing (and profit-reducing) technical requirements. It is significant that some of the most salient and memorable individual actors (Ezra Levin in the United States or H. Wilson Duval in Canada) were part of this group. At the same time, the majority of food manufacturing and fish processing companies in the United States refused to become involved in the private sector subsidization of FPC development due to the unfavorable ratio of cost to technological uncertainty.

Politicians on the other hand, saw in a publicly supported FPC program the possibility of simultaneously fulfilling three useful goals: provide the risk capital to develop a socially beneficial product for which the marketplace offered little incentive, utilize current, slack capacity in both fisheries resources and domestic labor, and fulfill desirable foreign policy objectives by serving as a visible, humanitarian source of foreign aid to the developing world.

Bureaucratic entrepreneurs were more than willing to encourage such inclinations among their political counterparts. They viewed FPC in terms of their particular agency's mission and the possibility of enhancing personal and organizational prestige and power (either by promoting or opposing the product). However, particularly in the case of the FPC proponents, positions were often adopted out of expediency and more as a result of intuitive judgment and wishful thinking than out of rational planning and analysis. Agency representatives, in their relationships with the legislative bodies and with each other, appeared more concerned with protecting or promoting their particular organizational position than with the impact that would result from developing and using the product.

Scientific leaders in the fields of nutrition, food science, and marine science were called upon at various times to study, advise, and endorse FPC development programs. Although scientific opinion considered FPC technology to be theoretically sound and potentially useful for the alleviation of one aspect of malnutrition as it was then understood, the individuals involved clearly were misinformed as to the product's economic viability. Particularly in the United States, scientific advisory groups were continually frustrated by the political and bureaucratic impediments that blocked FPC development and by the difficulty they encountered in obtaining reliable cost estimates. As a result, the normally influential impact of scientific advice was diluted by the myriad of political and bureaucratic obstacles with which it was forced to compete.

Finally, the *technologists,* chemists and engineers actually responsible for developing the product, conceived of FPC almost exclusively in terms of questions relating to its technical challenge. How could the technical obstacles be overcome? What kinds of specialized skills, capital equipment, and new technological innovation would be required to produce FPC? How long would the development process take? What would be the cost? Rarely were concerns voiced about the relationship between the design of the finished product and its likely consumers (with the exception of the Institute for Fisheries Development [IFOP]/Industrial Protein Group [CPI] project in Chile). Faced with the challenge of devising a product with universal application, technologists in particular became so involved in the process and its intermediate goals that development of the process became, in many respects, the *final* goal. FPC came to represent a technological "Mt. Everest," to be climbed because it was there.

Each of these entrepreneurial groups benefited from its involvement in FPC programs. Yet the fact that so many individuals and organizations, each with its own agenda of priorities, were vying to exert definitive influence over the course of FPC development inevitably meant that the varying interest groups were brought into conflict, and sometimes competition with one another. Ultimately, as the costs of pursuing these conflicting paths began to exceed the potential benefits, each group at some point decided to abandon the effort, dealing the FPC development effort a mortal blow in the process.

Critical Areas of Neglect in FPC Development

Enthusiasm for FPC led in many cases to gross oversimplification of the requirements for development and use. This in turn led to the miscalculation of costs and development time, illustrated by the history of the U.S. BCF project or the project at Quintero, Chile. Clearly, in some development efforts this kind of oversimplification can have beneficial effects by acting to obscure difficulties which, if fully revealed, might well cause a project's premature demise. But it can also lead to the distortion of appropriate objectives and real potential. In Morocco, for example, a market study, which would have revealed the total impracticality of manufacturing FPC for human consumption, was delayed in order to obtain funding for what was, in fact, to be a factory for the production of upgraded fish meal. In this case, the difficulties that were obscured were precisely those which made FPC an unlikely prospect for use as a nutrition intervention in low-income countries.

Almost entirely obscured by humanitarian rhetoric and actual FPC development were the real profiles of the populations whose nutritional needs were to be addressed. In Canada, the United States, and Sweden no specific populations were even considered as practical FPC target groups; the general idea in these countries was that the protein supplement would be shipped to developing countries as foreign aid. In Morocco, the projected development was ostensibly for school feeding programs as it was for a while in Chile. In all of these cases little concern was given to the design of the products in which FPC would be incorporated or to the needs and cultural habits of the people who would eat them. (Chile's case is the exception in terms of the Industrial Protein Group's work in developing specific products that would incorporate FPC and be used as milk substitutes and meat extenders.)

In general, it was tacitly understood that FPC was being developed for those who "needed" it most (that is, malnourished infants and young children and malnourished pregnant and nursing mothers). If developers had been really concerned, however, with those who needed a protein supplement, they would have been struck by the fact that such groups represent the most passive and powerless segments of low-income populations. Any serious attempt to define the nutritional needs of this segment of a population should have raised important questions as to the political and economic conditions that produce such nutritional deprivation. Such questions would surely have illuminated the inappropriateness of FPC as a protein nutrition intervention tool in these contexts.

In addition, any thorough examination of the intended consumers of FPC would have revealed that nutritional needs include not only quantitative dimensions (the need for enough food to provide sufficient energy for calorie expenditure and environmental and emotional stress) but also qualitative dimensions. These include the cultural patterns that determine food habits that exist in any group no matter how poor or politically powerless. The less industrialized or urbanized a population, the more conservative its food consumption system will be. Any attempt to make changes in that system requires a thorough knowledge of its rules. This kind of knowledge was rarely considered important in the course of FPC development.

Ironically, the food consumption habits that were most attended to in the development of FPC technology were those of the U.S. consumer, whose interests were allegedly represented by the FDA. It was in overall response to the demands of this agency that many of the cost-increasing aspects of FPC technology (such as desolventization and deboning) were added. The final product, greyish white, odorless, and tasteless, proved to be unnecessarily refined in the case of cultural groups that had an established preference for the taste and smell of fish in their diets. These groups include large Asian and African populations that might have been consumers of a less sophisticated FPC-type product.

In addition to ignoring the food habits of fish-eating populations, developers also failed to take into account the overall cultural significance of fish as a food source. Either revered or reviled, fish has played an important role in history and has strong cultural and social connotations attached to it. The decision to use fish as a raw material for a food supplement, no matter how bland, without respect for these connotations is evidence that the traditional and conservative nature of dietary systems and food preferences was generally unappreciated by FPC developers.

Today FPC is known not simply as a protein concentrate but as a *fish* protein concentrate. The identification of the product with fish is still inviolate and emphasis in discussion usually rests more upon the "fish" than on the "protein." The cultural connotations of fish bear ample evidence of the fact that food, while having a biological function, often has a much more important aesthetic and social function for the person eating it. Attempts to introduce new foods or ingredients which do not take these aesthetic and cultural functions into account have only a remote chance of success, especially in the least industrialized countries where food consumption habits are most tradition-bound.

Failure to Develop an Appropriate Technology

Although the nature of the protein malnutrition problem was defined in an incomplete and simplistic manner during the course of most FPC development, with little regard for the economic, social, and political factors that form the basis for its existence worldwide, it is still possible that a protein supplement might have proved useful for specific target groups in specific low-income countries. The product might have been useful if its production process were one that could be implemented by indigenous industries and manned by indigenous labor; if the country or locale in question had both a resource base and a fishing fleet adequate to ensure a steady supply; if the industry, rather than requiring the importation of foreign expertise, could have expanded employment possibilities for the country in question; if the product could have been priced in relation to the purchasing power of the poor; and finally, if the product had been designed in relation to the food habits and traditions of the populations in question.

FPC development efforts in almost all the countries we have considered did not satisfy these conditions. Nor was any effort made by FPC developers to examine the fish processing methods existing in developing countries for the purpose of streamlining them and/or designing some intermediate technology that might produce a product acceptable to the local population. Instead, a technology evolved that required a level of capital investment and technical expertise not readily available in developing countries.

If designed in developed countries, no matter what their ultimate goals, products like FPC (directed toward *macro-* rather than *micro*nutrient deficiencies) will in the course of development address a different set of concerns (in terms of processing, market structure, means of distribution, and so on) from those of developing countries. The transfer of expensive and elaborate technologies for the purpose of this type of nutritional intervention from the industrialized world to an area characterized by poverty and underdevelopment will probably fail; or if it does not, it will likely serve to perpetuate rather than eliminate the division between developers and recipients. The implementation of a high technology protein supplementation intervention may act as a palliative to publicized concern about protein-calorie malnutrition; it may by the same token make it easier to ignore the basic social problems that must be addressed if long-term equitable development is to occur.

Since the target populations of nutrition interventions are, by definition, impoverished, the cost of such projects, whether indigenous or imported, must be borne by the public sector, just as the cost of weapons procurement or environmental sanitation is paid for with public funds in both the developed and developing world. Public funds channeled toward the alleviation of protein-calorie malnutrition can, with adequate attention to existing economic and political realities, be much more effectively used than they were in the FPC development efforts we have studied. Innovations like FPC, while capable of affecting the *quality* of dietary intake, do so in a manner that is unlikely to affect malnourished groups because the cost of the technology puts the needed quantities of the product beyond the range of their effective economic demand without the application of massive government subsidies. Thus, it may well be that the most appropriate application of funds for protein-calorie interventions in low-income countries will actually have less to do with food technology than it will with related issues such as employment generation, land reform, agricultural development, or food pricing policies. Programs in these areas can more directly affect the nutritional status of malnourished groups, in terms of either absolute food supply or food purchasing power.

THE RELEVANCE OF FPC-TYPE APPLICATIONS FOR MALNUTRITION PROBLEMS IN THE DEVELOPING WORLD

After having examined the length and breadth of the FPC story, we are confronted with the realization that this product never possessed the potential to fulfill the principal mission for which it was nominally intended: the alleviation of malnutrition as it was then defined. This is not to suggest that FPC did not fulfill other, less emphasized objectives. The FPC programs did demonstrate that it was possible to utilize fishery stocks (if, in fact, such stocks existed) that might otherwise go unused. By the same token, FPC programs also proved capable of providing employment for idle labor capacity in the fishing and fish processing industry. Some have even argued that the product showed (and continues to show) the potential to return a profit under certain, limited conditions. But none of these justifications can begin to compensate for the fact that FPC failed as a means of alleviating protein malnutrition, which was, after all, its principal acknowledged purpose for existence.

The point here is that FPC programs did *not* fail as nutrition interventions primarily because of their myriad technical, political, economic, and human shortcomings, although all of these factors had a significant individual and synergistic impact. This conclusion may strike the reader as curious after having reviewed the evidence documented here. But the pattern that is clearly apparent, when all of this evidence is considered together, is that, by and large, FPC *never* reached the stage of a bona fide nutrition intervention in any of the cases presented. It did not do so because it was by its very design incapable of having a significant quantitative impact on the problem of protein malnutrition.

In reality, FPC represented a technological application conceived in the industrialized world in an environment and time period that were relatively free from constraints on capital, resources, and expertise. This product was to be applied in the developing world where all of these constraints existed in abundance. For all their technological sophistication and political and social attractiveness, the FPC programs failed to meet either of the principal criteria necessary for the successful implementation of a program designed to alleviate protein and/or calorie deficiencies: (1) FPC did not result in the increase of the real supply of protein or calories for those at nutritional risk; and (2) FPC did not increase the effective economic demand of those at risk for these nutrients. Since the technological nature of the program did not address itself to any of the economic and political factors that govern the distribution of effective economic demand for food, the hope was that the product would somehow increase the real supply of protein available to the chronically malnourished either by being extremely inexpensive to produce or by being distributed under substantial government subsidy. Ultimately, however, the problems of production, cost, and consumer and political acceptability precluded the possibility of FPC reaching its intended target populations in effective quantities.

Failing to meet either the increased supply or the increased demand criterion, FPC must be (and in most cases has been) relegated to the category of a sophisticated and capital-intensive technological solution in search of a problem to which it can be applied. The lesson to be drawn from the totality of the FPC experience is that enterprises directed at the alleviation of *macro-*

nutrient deficiencies that do not address at least one of the two primary criteria suggested here are structurally inappropriate and ultimately incapable of success. Whatever their intentions, they are to one degree or another only token palliatives offered by wealthy nations or local ruling elites in lieu of addressing the social and political factors underlying malnutrition. Attention to these problems is likely to require a politically and economically unacceptable reallocation of capital resources and development priorities. In our opinion there is little hope that "technological fixes" such as FPC can ever obviate the need to confront these fundamental concerns for social equity.

We are struck by the relative infrequency with which these lessons have been recognized and implemented by policymakers. As a result of this failure, we note with concern the continuing, useless proliferation of high technology nutrition interventions aimed specifically at protein-calorie malnutrition; like fish protein concentrate, they may be doomed to failure by their intrinsic conceptual myopia.

APPENDIX: ANNOTATED TECHNICAL REFERENCES

This section provides the reader with ready access to the wide range of scientific and technical literature in which FPC-related problems are discussed. We are attempting here to identify the most convenient entry points to some of the most important collections of information on selected issues and to provide leads to reliable and exhaustive sources of specific data. The first part contains reference material directing the reader to documents covering a wide range of general issues; in the second we have selected references to publications addressing specific technical problems.

CONTENTS

1. General Reference Material

1.1 BIBLIOGRAPHIES

Fish Protein Concentrate: A Comprehensive Bibliography. Library of Congress, Washington, D.C., 1970.

This bibliography was compiled by the Special Bibliographical Section, Science and Technology Division of the U.S. Library of Congress. In compiling the bibliography, the Library reviewed literature published from 1940 through 1969. The compilation contains 308 entries and an author and a subject index. Each entry is annotated, giving the most important highlights of the investigation. The bibliography is divided into the following headings: (1) General Aspect, (2) Processing Methods, (3) Chemical Composition, (4) Nutritional Studies, and (5) Acceptability and Use in Foods.

Fish Protein Concentrates (A Bibliography with Abstracts). Compiled by E. A. Harrison. National Technical Information Service, Springfield, Va. January 1975.

This 122-page bibliography covers the period between 1964 and October 1974; it contains 117 selected abstracts, including reports on the processing, nutritional value, storage, preparation, utilization, cost analysis, and marketing of FPC.

Fish Protein Concentrate (FPC) Program Review and Plan. Northrop Services, Inc., Subsidiary of the Northrop Corporation, 1700 North Lynn Street, Suite 1100, Arlington, Va. 22209. 1974.

This document represents a review of developments that occurred in connection with the U.S. Government-sponsored fish protein concentrate project under the aegis of the U.S. Bureau of Commercial Fisheries (BCF), later the National Marine Fisheries Service (NMFS). The comprehensive review of the program was contracted by the NMFS with Northrop Services, Inc., and resulted in a two-volume compilation of facts, cost estimates, and bibliographic entries describing U.S. Government efforts and achievements in this field from the project's inception to May 1972.

1.2 REVIEW ARTICLES

By far the most complete review is still the article by Roland Finch, "Fish Protein for Human Foods," *CRC Critical Reviews in Food Technology.* December 1970, pp. 519-580. The article is articulated in the following way: Fish Protein: (1) Fish Protein in Fish, (2) The Supply of Fish and Fish Protein and (3) Increasing the Supply of Fish Protein. *Fish Protein Concentrate* (FPC): (1) Definition of FPC, (2) FPC Processes, (3) Safety of FPC, (4) Composition and Nutritive Value of FPC, (5) FPC as a Protein Supplement in Human Foods, (6) Future Development of FPC, and (7) Conclusion. The article is based upon 288 references.

Another, shorter article is by E. R. Pariser, Proteins of Aquatic Origin as Foods for Human Consumption, Advances in Food Research, Academy Press, New York and London. *20:* 189 to 216, 1973. This contribution, with 68 references, deals with the aquatic protein resources; the volume of production; the trade in proteins of aquatic origin; the consumption of protein products of aquatic origin; consumer preference and demand; the husbanding of the resource; the trends in the utilization of aquatic protein resources; and the cultural, bureaucratic, and environmental obstacles to the greater use of the resource.

1.3 PROCEEDINGS OF CONFERENCES

International Congress on Fish in Nutrition, Washington, D.C. 1961. Edited by Eirik Heen and Rudolf Kreuzer, FAO, Rome. Fishing News (Books) Ltd., Ludgate House, 110 Fleet Street, London, E.C. 4, England. 1962.
This 447-page book contains over 70 original papers prepared by 150 authors, an excellent index, and French and Spanish translations of article abstracts. The proceedings are subdivided into the following major parts: (1) The Role of Fish in World Nutrition, (2) Chemical Components of Fish and Their Changes Under Treatment, (3) Contribution of Fish and Fish Products to National Diets, (4) Fish and Fishery Products in Animal Nutrition, and (5) Demand for Fish as Human Food and Possibilities for Increased Consumption.

CONFERENCE on FISH Protein Concentrate. Canadian Fisheries Reports No. 10. July 1968.
This conference was held under the general chairmanship of Dr. A. W. H. Needler, federal Deputy Minister of Fisheries, and brought together over 200 scientists, businessmen, national and international government personnel, and industrialists. Twenty-four original papers were presented under the following general headings: (1) The World Protein Situation, (2) Canadian Raw Material Resources, (3) Processing Technology, (4) Standards of FPC, (5) Development of the Industry, (6) Markets for FPC, and (7) Addenda: FPC in Foods and Feeding Trials on Undernourished Children with Fish Protein Concentrates.

Production of Fish Protein Concentrate. Report and Proceedings of the Joint UNIDO/FAO Expert Group Meeting. Rabat, Morocco, 8-12 December, 1969. United Nations Industrial Development Organization, Vienna. U.N., New York. Sales No.E.72 II. B.1, 1972.
This 170-page collection of 17 papers by 26 authors contains technical papers, ranging from topics on nutrition, quality control, and analytical methods for FPC to market and engineering aspects of FPC development.

Factors Affecting the Use of Fish Protein Concentrates. Report on an FAO symposium on the significance of fundamental research in the utilization of fish. *World Fish. Abstr. 15* (4): 43-44. 1964.

The Economics, Marketing, and Technology of Fish Protein Concentrate. Edited by Steven R. Tannenbaum, Bruce R. Stillings, and Nevin S. Scrimshaw. The MIT Press, Cambridge, Ma., and London, England. 1974.
This volume of 500 pages containing 27 papers is based on presentations made at the International Conference on Fish Protein Concentrate that was held at MIT on 6-8 June 1972. According to the book's preface, the contents of the volume are derived from the conference but do not cover all the papers given at the meeting. The presentations are grouped under the following headings: (1) Resources, (2) Processing, (3) Nutrition, (4) Utilization, and (5) Economics. The book contains a good subject index.

1.4 HEARINGS BEFORE THE U.S. SENATE

These hearings contain useful overall information on the project conducted by the U.S. Government. The statement by D. L. McKernan, Director of the Bureau of Commercial Fisheries, U.S. Department of the Interior, hearing before the Sub-Committee on Merchant Marine and Fisheries of the Committee on Commerce (U.S. Senate 88th Cong., 14 August 1969, Serial No. 60, U.S. Government Printing Office, Washington, D.C.) contains on pages 16-80 a briefing statement that deals in detail with the rationale, program, and progress of the BCF project. McKernan's statement is supplemented by one made by E. R. Pariser on pages 87-91, largely dealing with FPC processing methods.

The other hearings were conducted on 25 April and 16-17 May 1966, before the Committee on Commerce (U.S. Senate 89th Cong., sess. on S.2720, Serial No. 89-64. U.S. Government Printing Office, Washington, D.C.) and contain a number of statements by witnesses, including one by E. R. Pariser on pages 30-61 on the research and projected costs of the FPC project.

2. Specific Reference Material

2.1 RAW MATERIAL

This section contains references to articles in which the potential aquatic resources that are particularly suitable as raw materials for the manufacture of fish protein concentrates are discussed. In general, the series of articles in *Fish in Nutrition* (1.3) is a good introduction to this area, although the collection appeared fifteen years ago. A more recent but more restricted series of papers on the subject can be found in *The Economics, Marketing and Technology of Fish Protein Concentrate,* edited by S. T. Tannenbaum, which was published by The MIT Press in 1974. On the subject of raw material composition and properties, G. Borgstrom's book on *Fish as Food,* Academic Press, 1962, gives most of the basic and important gross information in this field.

2.1.1 General References

1966 Chapman, W. M. Ocean fisheries, status and outlook. In *Transactions, 2nd Annual Conference Marine Technology Society.* Washington, D.C., p. 15.

1967 The President's Science Advisory Committee. Report of the Panel on the World Food Supply. The Fishery Potential of the Ocean. U. S. Government Printing Office.

1968 Schaefer, M. B. and D. L. Alverson. World fish potentials. In *The Fu-*

ture of the Fishing Industry of the United States. University of Washington Press, p. 81.

1974 Juhl, R. Economics of Gulf of Mexico industrial and foodfish trawlers. *Marine Fish. Review 36* (11): 39–42.

1975 Bardach, J. and E. R. Pariser. Aquatic proteins. In *Protein Resources and Technology: Status and Research Needs,* principal investigators N. S. Scrimshaw and D. I. C. Wang. NSF-RA-T-75-037.

2.1.2 Raw Material Availability

1968 Cunningham, A. Fillet trimmings as a potential raw material for FPC. Conference on FPC. Canadian Fisheries Reports No. 10.

1968 Martin, W. R. Fish resources and their potential utilization for FPC. Conference on FPC. Canadian Fisheries Reports No. 10.

1968 Meagher, B. The potential use of canadian raw material resources for the manufacture of FPC. Conference on FPC. Canadian Fisheries Reports No. 10.

1969 Roels, O. A. Marine protein. *Chemical Engineering Progr. 65* (9):27.

1972 Kreuzer, R. Potential resources for the industrial production of FPC. In *Production of Fish Protein Concentrate.* Report and proceedings of the Joint UNIDO/FAO Expert Group Meeting. Rabat, Morocco, 8–12 December 1969. United Nations Industrial Development Organization, Vienna. U.N., New York. Sales No. E.72 II. B.1.

2.1.3 Raw Material Composition

1959 Love, R. M., J. A. Lovern, and N. R. Jones. The chemical composition of fish tissues. D.S.I.R. Food Investigation. Special Report No. 69. H.M.S.O. London.

1961 Stansby, M. E. Proximate composition of fish. In *Fish in Nutrition,* eds. E. Heen and R. Kreuzer. London: Fishing News (Books) LTD.

1962 Bramsted, F. A. L. Amino acid composition of fresh fish and the influence of storage and processing. In *Fish in Nutrition,* eds. E. Heen and R. Kreuzer. London: Fishing News (Books) LTD.

1963 Stansby, M. E., and H. S. Olcott. Composition of fish. In *Industrial Fishery Technology,* ed. M. E. Stansby. New York: Reinhold Publishers.

1965 Thompson, M. H. and R. N. Farragut. Amino acid composition of the alewife (Alosa Pseudo-Harengus). *Fishery Industry Research* (3):47–53.

2.2 FISH PROTEIN CONCENTRATE MANUFACTURING METHODS

The most common processing methods have so far been based upon the use of an organic solvent with which the comminuted raw material is simultaneously defatted and dehydrated. The solvent is recovered by distillation and suitably deodorized and then reintroduced into the defatting/dehydration stage of the manufacturing process. Water and fats extracted by the solvent are by-products. Other, more promising methods, utilizing enzymes, are being developed. Their advantage appears to reside in being less expensive and leading to products with better functional properties.

Roland A. Finch's article in CRC (1.2) is still the most complete review with, by far, the most extensive list of references. The article provides the largest number of pointers to FPC manufacturing methods and specific process-related problems.

This section consists of three groups of references: the first fifteen references point to articles of general interest. They also include papers on novel processing methods and provide entry to publications dealing with the effects of processing conditions upon the properties of the final product. The second group of references contains publications concerning some of the major FPC processing methods discussed in the book. The third group is made up of a selection of thirty-three patents relating to the manufacture of FPC that were granted in seven countries during the last half century.

2.2.1 General References

1962 Lovern, J. A. The lipids of fish and changes occurring in them during processing and storage. In *Fish in Nutrition,* eds. E. Heen and R. Kreuzer. London: Fishing News (Books) Ltd.

1963 Pariser, E. R., and E. Odland. MIT–UNICEF Studies on the Production of Fish Protein Concentrate for Human Consumption. *Commercial Fisheries Review 25* (10).

1965 Miller, D. S. The nutritive value of fish proteins. *J. Science Food Agr. 7:*337.

1968 Pariser, E. R. Food potential of the seas and its significance to food processors. Proceedings, Third Biennial Food Forum. 14–16 October 1968. Washington, D. C.: Dairy and Food Industry Supply Association.

1970 Tannenbaum, S. R., R. P. Bates, and L. Bradfield. Solubilization of fish protein concentrate. II. Utilization of the alkaline process product. *Food Technology 24* (99).

1970 Tannenbaum, S. R., M. Aherne, and R. P. Bates. Solubilization of fish protein concentrate. I. An alkaline process. *Food Technology 24* (96).

1970 Spinelli, J. and B. Koury. Phosphate complexes of soluble fish proteins, their formation and possible use. *J. of Agriculture and Food Chemistry 18* (284).

1972 Roels, O. Fish protein concentrate—history and trends in production. Joint UNIDO/FAO Expert Group Meeting 8–12 December 1969. Production of FPC. UN Sales No. E.72II.B.1.

1973 Miyauchi, D., G. Kudo, and M. Patashnik. Surimi—a semi-processed wet fish protein. *Fisheries Review 35* (12): 7–9.

1973 Groninger, H. S., Jr. Preparation and properties of succinylated fish myofibrillar proteins. *J. of Agriculture and Food Chemistry 21* (6): 978–981.

1974 Dubrow, D. L., A. Kramer, and A. D. McPhee. Effect of temperature on lipid extraction and functional properties of fish protein concentrate. *Journal of Food Science 38* (6):1012–1015.

1974 Rasekh, J. Effect of freeze drying, spray drying and stabilizers on functional characteristics of fish protein concentrate. *Journal Milk Food Technology 37* (2):78–82.

1974 Valle-Riestra, J. F. Food processing with chlorinated solvents. *Food Technology 28* (2):25–32.

1974 Pelroy, G., J. Spinelli, R. Miller, D. Wieg, and L. Lehman. Stabilized press cake for FPC (fish protein concentrate) manufacture. *Food Technology 28* (3):64–82.

1974 Spinelli, J. Aqueous solvent processes. In *The Economics, Marketing, and Technology of Fish Protein Concentrate,* eds. Steven R. Tannenbaum, Bruce R. Stillings, and Nevin S. Scrimshaw. Cambridge, Ma.: The MIT Press.

2.2.2 References to Selected Processing Methods

Dabsch-UNICEF

1963 Allen, L. E. Fish flour production in Chile. *Fish. News International 2:* 106–110.

1963 Pariser, E. R. and E. Odland. MIT-UNICEF Studies on the Production of Fish Protein Concentrate for Human Consumption. *Commercial Fishery Review 25* (10):6–13.

Fishery Research Board of Canada

1957 Guttman, A., and F. A. Vandenheuvel. The production of edible fish protein (fish flour) from cod and haddock. *J. Fish. Res. Bd. Canada, Progress Report, Atlantic Coast 67:* 29–31.

1959 Dambergs, N. Extracts of fish muscle. 2. Solvent-water ratio in extraction of fat and water solubles. *J. Fish. Res. Bd. Canada 16:* 63–71.

1967 Guttman, A., N. Dambergs, and P. M. Jangaard. Preservation of raw material for fish protein concentrate with isopropyl alcohol. *J. Fish. Res. Bd. Canada 24:* 895.

1968 Idler, D. R. The development of the Halifax process. Conference on Fish Protein Concentrate. Canadian Fishery Reports No. 10.

1969 Dambergs, N. Isopropanol-water azeotrope as solvent in the production of FPC from herring (Clupea Harengus). *Journal Fisheries Research Bd. Canada 26:* 1923.

1969 Dambergs, N. Isopropyl alcohol-water mixtures for the production of FPC from Atlantic Herring (Clupea Harengus). *J. Fish. Res. Bd. Canada 27:* 1919.

1972 Idler, D. R. The Halifax isopropanol process for the manufacture of FPC. In *Production of Fish Protein Concentrate.* Report and proceedings of the Joint UNIDO/FAO Expert Group Meeting. Rabat, Morocco, 8–12 December 1969. United Nations Industrial Development Organization, Vienna. U.N., New York. Sales No. E.72 II. B.1.

1974 Regier, L. W. Canadian fish protein concentrate and development. In *The Economics, Marketing and Technology of Fish Protein Concentrate,* eds. Steven R. Tannenbaum, Bruce R. Stillings, and Nevin S. Scrimshaw. Cambridge, Ma.: The MIT Press.

Nabisco-Astra Nutrition

1970 Hallgren, B. and L. B. Sjoberg. Research and product development. Nabisco-Astra Nutrition Corp., New York.

1976 Knutson, Leif, (ed.) The Astra FPC-process. *Astra Protein Letter 4* (2).

U.S. Bureau of Commercial Fisheries (Now the National Marine Fisheries Service)

1966 U.S. Department of the Interior. Marine protein concentrate. Fishery Leaflet 584. Washington, D.C.

1969 Brown, N. L., and H. Miller, Jr. Experimental production of fish protein concentrate (FPC) from Mediterranean sardines. *Commercial Fishery Review 31* (10):30–33.

1969 Dubrow, D., and O. Hammerle. Holding raw fish (Red Hake) in isopropyl for fish protein concentrate production. *Food Technology 23:* 254.

1969 Finch, R. The U.S. fish protein concentrate program. *Commercial Fish. Review.*

1970 Goldmintz, D., and J. C. Hull. Bacteriological aspects of fish protein concentrate. *Dev. Ind. Microbiology 11:* 335.

1974 Hale, M. B. Using enzymes to make fish protein concentrates. *Mar. Fishery Review 36* (2):15–18.

1974 Dubrow, D. L., A. Kramer, and A. D. McPhee. Effect of temperature on lipid extraction and functional properties of fish protein concentrate (FPC). *Journal Food Science 38* (6):1012–1015.

VioBin

1951 Levin, E. and R. K. Finn. A process for dehydrating and defatting tissues at low temperature. *Chemical Engineering Progr. 51:* 223–226.

1959 Fish flour and meal by azeotropic solvent processing. *Food Technology 13* (2):132–135.

Wiking Eiweiss

1952 Shenstone, F. S. Egg white and yolk substitute. *Wiking Eiweiss Food Pres. Quart. 13* (3):45.

2.2.3 References to Selected Manufacturing Patents

1933 Ash, C. S. U.S. Patent 1,934,677.

1943 Hiltner, P. P., H. Metzner, and H. Peters. German Patent 733,896.

1950 Levin, E. Dried defatted enzymic material. U.S. Patent 2,503,313.

1950 Worsham, E. M. Levin, E. Simultaneous defatting and dehydrating of fatty substances. U.S. Patent 2,503,312.

1951 Levin, E. and E. M. Worsham. U.S. Patent 2,539,544.

1952 Levin, E. U.S. Patent 2,619,425.

1953 Greenfield, C. U.S. Patent 2,651, 647.

1955 Brody, J. U.S. Patent 2,717,835.

1957 Mohler, K., and R. Vogel. German Patent 1,005,360.

1959 Vogel, R., and K. Mohler. U.S. Patent 2,875,061.

1959 Bertullo, V. H., and F. P. Hettich. U.S. Patent 3,000,789.

1961 Levin, E. U.S. Patent 2,972,542.

1963 Moshy, R. J. Canadian Patent 663,556.

1963 Verrando, C. U.S. Patent 3,064,018.

1963 Cavanagh, J. C., and E. G. Inman. German Patent 1,138,308.

1963 Moshy, R. J. Canadian Patent 663, 559.

1963 Hasdenteufel, J. B. French Patent 1,534,769.

1963 The Pillsbury Corp. British Patent 940,309.

1965 Jeffreys, G. A., and A. J. Krell. U.S. Patent 3,170,794.

1966 Keys, C. W., and W. W. Meinke. U.S. Patent 3,249,442.

1966 Whaley, W. M., and R. J. Moshy. U.S. Patent 3,252,962.

1967 Somlai, I. French Patent 492,228.

1967 Libenson, C. M., and I. Pirosky. French Patent 1,502,113.

1967 Bohrer, B. H. U.S. Patent 3,298,109.

1967 Nestlé, S. A. Netherlands Patent Appl. 6,612,977.

1968 Somlai, I. British Patent 1, 106, 676.

1968 Linson, E. V. Edible marine protein concentrate. S. African Patent 6,705,900.

1969 Saloman, A. British Patent 1,155,047.

1969 Blaw Knox Co. British Patent 1,167,673.

1969 Levin, E. British Patent 1,127,329

1969 Lum, K. C. British Patent 1,157,415.

1970 Bertullo, V. U.S. Patent 3,516,349.

1974 Spinelli, J., and B. J. Koury. U.S. Patent 3,826,848.

2.3 THE FINISHED PRODUCT: COMPOSITION AND PROPERTIES

Composition and properties of the finished product determine whether and to what extent the FPC is marketable and useful as a dietary protein supplement. In Canada and the United States, precise standards exist, defined and described by the U.S. Food and Drug Administration and the Canadian Food and Drug Directorate, concerning the protein, lipid, ash, moisture, and mineral contents that are required and/or permissible if the FPC is to be manufactured and/or sold in these countries. Maximum microbiological loads and residual solvent levels are also prescribed in these official documents. The "functional" properties of the product concern its physical and chemical characteristics that determine importantly how successfully the FPC can be mixed with other food ingredients. "Organoleptic" properties describe the product's taste, odor, and mouth feel and have an important bearing upon the FPC's acceptability and marketability.

The references on product composition and properties that were selected for this small bibliography focus largely upon the nutritive value of FPC; the entries are grouped in three major categories. In the first group are titles of publications describing FPC in general; in the second group are important papers to introduce the reader to problems concerning the residual lipid, solvent, volatile, and fluoride residues in the product as well as publications concerning the nutrient value of FPC. The third group of references is subdivided into two sections: one, dealing with more specifically nutrition and safety oriented topics; the other, with concerns relating to the organoleptic and functional properties of the protein concentrate.

2.3.1 General References

1962 Pariser, E. R. Fish protein concentrate—A high quality animal protein. *Commercial Fishery Review 24* (5).

1964 Power, H. E. Characteristics and nutritional value of various fish protein concentrates. *Journal Fishery Res. Bd. Canada 21:*1486.

1966 Lovern, J. A. General outlook for edible fish protein concentrates. In *World Protein Resources, Advances in Chemistry* Series No. 57. Washington, D.C.: Am. Chemical Society.

1972 Chichester, C. O., F. Monckeberg, and E. Yanez. The nutrition effectiveness and acceptability of FPC. In *Production of Fish Protein Concentrate.* Report and proceedings of the Joint UNIDO/FAO Expert Group Meeting. Rabat, Morocco, 8–12 December 1969. United Nations Industrial Development Organization, Vienna. U.N., New York. Sales No. E.72 II. B.1.

2.3.2 Finished Product Composition

1963 Morrison, A. B. and Z. I. Sabry. Factors influencing the nutritional value of fish flour. II. Availability of lysine and sulfur amino acids. *Can. J. Biochemical Physiology 41:*649.

1966 U.S. Department of the Interior. Marine protein concentrate. Fishery Leaflet 584. Washington, D.C.

1967 Medwadowski, B. F., J. Vanderveen, and H. S. Olcott. Nature of the residual lipids in fish protein concentrate (FPC). *J. Food Science 32:* 361.

1967 Wick, E. L., E. Underiner, and E. Paneras. Volatile constituents of fish protein concentrate. *Journal Food Science 32:*365.

1968 Medwadowski, B. F., J. Vanderveen, and H. S. Olcott. Nature of residual lipids in menhaden FPC. *J. Am. Oil Chem. Soc. 45:*709.

1968 Moorjani, M. N., N. L. Lahiry, R. B. Nair, A. N. Upadhye, and S. V. Rae. Nutritional value of fish flour. I. Effects of storage on sardine meal prior to extraction. *Food Technology 22:*1557.

1968 Ackman, R. G., and P. H. Odense. Retention of isopropyl alcohol in fish protein concentrates. *J. Fish. Res. Bd. Canada 25:*804.

1969 Stillings, B. R., O. Hammerle, and D. G. Snyder. Sequence in limiting amino acids in fish protein concentrate produced by isopropyl alcohol extraction of red hake (Urophycis chuss). *Journal Nutrition 97:*70.

1970 Dubrow, D. L., E. R. Pariser, N. L. Brown, and H. Miller. FPC's quality virtually the same as its raw material quality. *Com. Fish. Rev.,* pp. 25–31.

1970 Weinberg, B. The current status of fish protein concentrate. Paper presented to the conference on Marine Sciences. University of Prince Edward Island, Charlottetown, P.E.I. 16–18 August.

1970 Ke, P. J., H. E. Power, and L. W. Regier. Fluoride content of fish protein concentrate and raw fish. *J. Sci. Food Agric. 21:*108.

1971 Dubrow, D. L., N. L. Brown, E. R. Pariser, H. Miller, V. D. Sidwell, and M. E. Ambrose. Effect of ice storage on the chemical and nutritive properties of solvent extracted whole fish—red hake (Urophycis chuss). *Fishery Bull. 69* (1):145–150.

1974 Smith, V. J., J. S. Linn, and H. S. Olcott. The residual lipids of fish protein concentrates. *Mar. Fish. Serv. Fish. Bull. 72* (3):845–847.

2.3.3 Finished Product Properties

Nutrition and Safety

1960 Carpenter, K. J. The estimation of available lysine in animal protein foods. *Biochem. Journal 77:*604–610.

1960 Campbell, J. A. Evaluation of protein in foods for regulatory purposes. *Agric. Food Chem. 8:*323–327.

1961 Jaffe, W. G. New studies on the nutritive effect of fish flour and its supplementary action on wheat flour and bread *Arch. Venezol. Nutr. 14* (19).

1962 Graham, G. G., J. M. Baerth, and A. Cordano. Evaluation of fish flour in the treatment of infant malnutrition. In *Fish in Nutrition,* eds. E. Heem and R. Kreuzer. London: Fishing News (Books) Ltd.

1962 Morrison, A. B., Z. I. Sabry, and E. J. Middleton. Factors affecting the nutritional value of fish flour. I. Effects of extraction with chloroform or ethylene dichloride. *J. Nutr. 77:*97.

1963 Doraiswamy, T. R., S. R. Shurpalekar, M. N. Moorjani, N. I., Lahiry, A. N. Sankaran, M. Swaminathan, A. Sreenivasan, and M. Subrahmanyan. The effect of a supplementary protein food containing fish flour, groundnut flour, and bengal gram flour, and fortified with vitamins on the growth and nutritional status of children. *Indian J. Pediat. 30:*266.

1964 Power, H. E. Characteristics and nutritional value of various fish protein concentrates. *J. Fish. Res. Bd. Canada 2:*1489–1509.

1964 Pretorius, P. J., and E. S. Wehmeyer. An assessment of nutritive value of fish flour in the treatment of convalescent kwashiorkor patients. *Amer. J. Clin. Nut. 14:*147.

1964 Lee, K. Y., M. S. Sin, M. H. Kim, E. S. Lee, B. K. Cho, and J. A. Lewis. Effect of protein supplementation by fish flour weaning period. *Yansi Med. J. 5.*

1964 Hadjimarkos, D. M. Fluoride in fish-flour effect on teeth. *J. Pediat. 65:* 782.

1965 Pereira, S., T. Isaac, B. Tewarson, and M. E. Dumm. Processed fish protein in the treatment of kwashiorkor. *Indian J. Med. Res. 53:*651.

1965 Morrison, A. B., and I. C. Munro. Factors affecting the nutritional value of fish flour. IV. Reaction between 1.2 dichlorethane and protein. *Can. J. Biochem. 43* (33).

1967 Stillings, B. R. Nutritional evaluation of fish protein concentrate. Activities Report, Research and Dev. Assoc., Inc., Natick, Ma. *19:*109.

1967 U.S., Federal Register. Whole fish protein concentrate. 2 February 1967, p. 1173.

1967 Yanez, E., I. Barja, F. Monckeberg, A. Maccioni, and G. Donoso. The FPC story. Quintero fish protein concentrate. Protein quality and use in foods. *Food Technology 21:*1604.

1967 Munro, I. C., and A. B. Morrison. Factors influencing the nutritional value of fish flour. V. Chlorocholine chloride, a toxic material in samples extracted with 1.2 dichlorethylene. *Can. J. Biochem. 45:*1049.

1967 Munro, I. C., and A. B. Morrison. Toxicity of 1.2 dichloroethane extracted FPC. *Can. J. Biochem 45:*1779.

1968 Power, H. E. Nutritional characteristics and projections on production costs of products prepared by the Halifax process. Conference on FPC. Canadian Fishery Reports No. 10.

1968 Morrison, A. B. Safety of FPC. Conference on FPC. Canadian Fishery Reports No. 10.

1969 Munro, I. C., A. B. Morrison, and M. Myer. Fish protein concentrate as a supplement to cereal diets. *J. Am. Diet. Assoc. 54:*398.

1969 Grütner, R., and K. H. Schäfer. Biological value of fish protein in nutritional experiments with premature infants and animals. *Klin. Wochenschr. 37:*225.

1969 Sen, D. P., T. S. Syanarayana Rao, S. B. Kadkol, M. A. Krishnaswamy, S. Venkata Rao., and N. L. Lahiry. Fish protein concentrate from Bombay duck (Harpodon nehereus) fish: effect of processing variables on the nutritional and organoleptic qualities. *Food Technology 23:*683.

1969 Ershoff, B. H., and P. G. Rucker. Nutritive value of 1.2 dichloroethane extracted fish protein concentrate. *J. Food Science 34:*355.

1969 Beasley, T. M., and C. L. Osterberg. Natural and artificial radionuclides in seafoods and marine protein concentrates. *Nature 211:*1207.

1970 Ershoff, B. H., J. B. Lol, and S. Bernick. Beneficial effects of fish protein concentrate on increment in body weight and microscopic appearance of the tibia and alveolar bone of rats fed a wheat flour-containing ration. *J. of Dental Res. 49* (3):581-588.

1970 U.S., *Federal Register.* Whole fish protein concentrate. 4 August 1970, p. 12390.

1970 Zipkin, I., and S. M. Zukas. Biological availability of the fluoride of fish protein concentrate in the rat. *J. Nutr. 100:*293.

1970 Spencer, H., J. Fowler, M. J. Kulka, and J. Samacsen. Availability of protein and minerals from fish protein concentrate in man. Third Int. Congress of Food Sci. and Technology, Washington, D.C., Abstract 195.

1970 Hadjimarkos, D. M. Fluoride content of fish protein concentrate. *Amer. J. Nutr. 23:*684.

1971 Friedman, L., O. G. Glaser, N. L. Brown, and E. R. Pariser. The wholesomeness of fish protein concentrate. A new approach to the evaluation of food safety. *Toxic. and Appl. Pharmac. 18:*329-352.

1973 Hauschild, A. H. W., H. Pivnick, and L. W. Regier. Inability to detect spores of Clostridium Botulinum in fish protein concentrates. *J. Fish. Res. Bd. Can. 30* (11):1760-1762.

1973 Regier, L. W. Mercury removal from FPC. *J. Fish. Res. Bd. Canada 29* (12):1777-1779.

1973 Archer, M. C., B. R. Stillings, S. R. Tannenbaum, and D. C. Wang. Reduction in mercury content of FPC by enzymatic digestion. *J. Agric. Food Chem. 21* (6):1116-1117.

1973 Fox, W. E. Sanitary considerations in the production of FPC. In *Microbial Safety of Fishery Products,* ed. C. L. Chichester, pp. 125-129.

1973 Newberne, P. M., O. Glaser, and L. Friedman. Biologic adequacy of FPC in five generations of mice. *Nutrition Reports International* 7 (4): 181-192.

1973 Newberne, P. M., O. Glaser, and L. Friedman. Safety evaluation of FPC over five generations of rats. *Toxic. Appl. Pharmacol. 24* (1):133-141.

1973 Paskell, S. L., and D. Goldminz. Bacteriological aspects of FPC production from a large-scale experiment and demonstration plant. *Developments in Industrial Microbiology 14:*302-309.

1973 Stillings, B. R., H. R. Lagally, E. Zook, and I. Zipkin. Further studies on the availability of the fluoride in FPC. *J. Nutrition 103* (1):26-35.

1974 Stillings, B. R. Nutritional and safety characteristics of FPC. In *The Economics, Marketing and Technology of Fish Protein Concentrate,* eds. Steven R. Tannenbaum, Bruce R. Stillings, and Nevin S. Scrimshaw. Cambridge, Ma.: MIT Press.

1974 Scrimshaw, N. S. Evaluation of fish protein concentrate for infant and child feeding. In *The Economics, Marketing, and Economics of Fish Protein Concentrate,* eds. Steven R. Tannenbaum, Bruce R. Stillings, and Nevin S. Scrimshaw. Cambridge, Ma.: The MIT Press.

1974 Balerezo, E. A., F. S. Nury, R. Newsome, and D. Nelson. Physiological influence of dietary fish protein concentrate in rats. *J. Animal Science 33* (6):1332.

1974 Yu, T. C. Quality and nutritive value of fishery products. Oregon State University, Agric. Expt. Station, Corvallis, Or.

1974 Makdani, D. D., J. T. Huber, O. Mickelson, and W. G. Bergen. The influence of water fractionation on the nutritional value of fish protein concentrate. *Nutrition Rep. Int. 9* (5):305-317.

1974 Hoskins, F. H., and J. Loustaunau. Fish protein concentrates as a mineral nutrient source. *Food Technology 28* (3):58-62.

1975 Spencer, H., L. Kramer, D. Osis, and E. Wiatrowski. Excretion of retained fluoride in man. *J. Applied Physiology 38* (2):282-287.

Organoleptic and Functional Properties

1973 Maga, J. A., and K. Lorenz. Flavor quality of various grain flours and oilseed, milk and marine protein supplement. *J. Milk Food Technol. 36* (4):232-236.

1974 Richardson, T., and C. H. Amundson. Modification of functional properties of proteins for use in food systems. University of Wisconsin, Agric. Exp. Station, Madison, Wi.

1974 Spinelli, J., and B. J. Koury. Functional fish protein concentrates and isolates. U.S. Patent 382,6848.

1975 Chen, L. F., T. Richardson, and C. H. Amundson. Some functional properties of succinylated proteins from fish protein concentrate. *J. Milk Food Technol. 38* (2):89-93.

1975 Noguchi, M., M. Yamashita, S. Arai, and M. Fujimaki. On the bitter masking activity of a glutamic-acid-rich oligo-peptide fraction. *J. Food Science 40* (2):367-369.

2.4 USE AND ECONOMICS OF FPC

A great wealth of published material exists in the scientific literature describing the use of FPC in various food products. The comprehensive bibliography on FPC issued by the Library of Congress (1.1.1) contains a large number of references, but the reader is also advised to consult more recent publications by V. D. Sidwell in this regard. As to the economics of an FPC enterprise and the considerations involved in setting up a marketing strategy for the product, the literature is less extensive, but the detailed papers by Hammonds and Call (2.4.3) are recommended as informative and thought-provoking documents.

2.4.1 General References

1968 Holme, J. Potential utilization by the cereal products industry. Conference on FPC. *Can. Fishery Reports* No. 10.

1970 Hammonds, T. M., and D. L. Call. Utilization of protein ingredients in the U.S. food industry. Cornell University, Ithaca, N.Y. Dept. of Agric. Economics.

1970 Sidwell, V. D., B. R. Stillings, and G. M. Knobl. The fish protein concentrate story, 10. U.S. Bureau of Commercial Fisheries. FPC Nutritional quality and use in foods. *Food Technol. 24:*876.

1972 Kapsiotis, G. D. Utilization and quality control of FPC. In *Production of Fish Protein Concentrate.* Report and proceedings of the Joint UNIDO/FAO Expert Group Meeting. Rabat, Morocco, 8–12 December, 1969. United Nations Industrial Development Organization, Vienna. U.N., New York. Sales No. E. 72 II. B.1.

2.4.2 Use of FPC in Foods and Feeds

Use in Foods

1939 Larsen, L. The use of fishmeal in bread. *Tids Kjemi Bergvesen 19* (150):161.

1964 Rusoff, I. J., A. H. Goodman, J. Sommer, and S. M. Cantor. Protein fortification of doughnuts. *Food Technol. 18:*131.

1967 Sidwell, V. D. FPC in foods. Activities Report, Research and Dev. Assocs., Inc., Natick, Ma. *19:*118.

1972 Sidwell V. D., B. R. Stillings, and G. M. Knobl, Jr. Analysis, testing and uses of FPC. In *Production of Fish Protein Concentrate.* Report and proceedings of the Joint UNIDO/FAO Expert Group Meeting. Rabat, Morocco, 8–12 December 1969. United Nations Industrial Development Organization, Vienna. U.N., New York. Sales No. E. 72 II. B.1.

1973 Pariser, E. R. Proteins of aquatic origin as foods for human consumption. *Advances in Food Research 20:*189–216.

1973 Sidwell, V. D. Candy fortified with fish protein. *Manufacturing Confectioner,* pp. 44–48.

1974 Sidwell, V. D. Utilization of fish protein concentrate and acceptability. In *The Economics, Marketing, and Technology of Fish Protein Concentrate,* eds. Steven R. Tannenbaum, Bruce R. Stillings, and Nevin S. Scrimshaw. Cambridge, Ma: The MIT Press.

1974 Dreosti, G. R. Acceptability of fish protein concentrate products. In *The Economics, Marketing and Technology of Fish Protein Concentrate,* eds. Steven R. Tannenbaum, Bruce R. Stillings, and Nevin S. Scrimshaw. Cambridge, Ma.: MIT Press.

1974 Pratt, D. E., and E. F. Reber. Nutritional value of non-conventional foods. Purdue University, Agric. Exp. Sta., Lafayette, In.

1974 Young, V. S. Evaluation of the protein quality of fish protein concentrate for maintenance of adults. In *The Economics, Marketing, and Technology of Fish Protein Concentrate,* eds. Steven R. Tannenbaum, Bruce R. Stillings, and Nevin S. Scrimshaw. Cambridge, Ma.: The MIT Press.

1974 Yanez, E., D. Ballester, and F. Monckeberg. The enrichment of foods with fish protein concentrate. In *The Economics, Marketing and Technology of Fish Protein Concentrate,* eds. Steven R. Tannenbaum, Bruce R. Stillings, and Nevin S. Scrimshaw. Cambridge, Ma.: The MIT Press.

Use in Feeds

1967 Huber, J. T., and L. M. Slade. Fish flour as a protein source in calf milk replacer. *J. Dairy Sci. 50:*1296.

1974 Crawford, D. L., D. K. Law, and T. B. McKee. Nutritional characteristics of Oregon pellet rations containing meals of different fish species. *The Progressive Fish. Culturist 36* (1):3–7.

1975 Huber, J. T. Fish protein concentrate and fish meal in calf milk replacers. *J. Dairy Sci. 58* (3):441–447.

Economics of FPC

1970 Devanney, J. W., and G. Mahnken. The economics of fish protein concentrate. MIT, MIT Sea Grant Program Report, Cambridge, Ma.

1972 Devanney, J. W. The economics of FPC Proceedings Western Hemisphere Nutrition Congress III. New York: Futura Publishing Co.

1974 Crutchfield, J. A. and R. Deacon. The economics of fish protein concentrate. In *The Economics, Marketing and Technology of Fish Protein Concentrate,* eds. Steven R. Tannenbaum, Bruce R. Stillings, and Nevin S. Scrimshaw. Cambridge, Ma.: The MIT Press.

1974 Pontecorvo, G. Marketing and economics: Peru. Fishmeal and FPC. In *The Economics, Marketing and Technology of Fish Protein Concentrate,* eds. Steven R. Tannenbaum, Bruce R. Stillings, and Nevin S. Scrimshaw. Cambridge, Ma.: The MIT Press.

1974 Pariser, E. R. Factors affecting the economics of an FPC manufacturing operation. In *The Economics, Marketing and Technology of Fish Protein Concentrate,* eds. Steven R. Tannenbaum, Bruce R. Stillings, and Nevin S. Scrimshaw. Cambridge, Ma.: The MIT Press.

1974 Kreuzer, R. Some problems in marketing FPC. In *The Economics, Marketing and Technology of Fish Protein Concentrate,* eds. Steven R. Tannenbaum, Bruce R. Stillings, and Nevin S. Scrimshaw. Cambridge, Ma.: The MIT Press.

1974 Champagne, J. R. Some aspects of the fish protein market in industrialized countries: a view from private industry. In *The Economics, Marketing and Technology of Fish Protein Concentrate,* eds. Steven R. Tannenbaum, Bruce R. Stillings, and Nevin S. Scrimshaw. Cambridge, Ma.: The MIT Press.

NOTE TO INTRODUCTION

1. FPC—known until 1962 as fish flour—differs from fish meal in that FPC denotes a product with a higher protein content (above 75 percent, frequently above 85 percent), a lower lipid content (below 1 percent, frequently in the vicinity of 0.1 percent), and a lighter color than that typical for fish meal. An important further distinction is the fact that FPC is manufactured from edible grade raw material by methods satisfying food processing standards, whereas fish meal is prepared from any available fish by crude production methods utilizing equipment not usually suitable for the production of food for human use.

NOTES TO CHAPTER 1

1. D. Bohlen, *Die Bedeutung der Fischerei für die Antike Wirtschaft* (Hans Christians Druckerei und Verlag) (Hamburg: Kleine Theaterstrasse, 1937), p. 11.

2. *Ibid.*

3. W. Radcliffe, *Fishing from the earliest times,* 2nd ed. (New York: E. P. Dutton and Co., 1926), p. 5.

4. *Ibid.,* p. 375.

5. Athenaeus, *The Deipnosophists,* trans. by Ch. B. Gulick, The Loeb Classical Library, Vol. I (Cambridge, Mass.: Harvard University Press, 1969), Book

I, p. 57. Athenaeus lived from about A.D. 170 to A.D. 230 in Rome but was a Greek national native of Naucratis in Egypt. The "Deipnosophists," or "lunch-time philosophers," "dinner-table gossips" is a collection of countless anecdotes and literary quotations by Athenaeus who pretends that he is telling a friend about a banquet at the home of a scholar to which many literary figures have been invited. This gives Athenaeus an excuse to show off, almost at random, his erudition and detailed information on the behavior of cooks, exotic dishes, table manners, recipes, menu-cards, quoting 800 writers and pieces of writing, many of which would otherwise have remained unknown to us.

6. Aristotle, *History of Animals,* trans. by A. L. Peck, the Loeb Classical Library (Cambridge, Mass.: Harvard University Press).

7. Athenaeus, *The Deipnosophists,* Book VII, p. 308.

8. Examples in this paragraph are from Otto Keller, *Die Antike Tierwelt,* Georg Olms Verlagsbuchhandlung, Vol. II (Hildesheim), 1963.

9. E. R. Goodenough, *Jewish symbols in the Greco-Roman Period,* Bollingen Series 37 (New York: Pantheon Books, 1935), Vol. 5, p. 33.

10. I. Scheftelowitz, *Die Alt-Persische Religion und das Judentum* (Giessen: A. Toeppelmann, 1920).

11. Radcliffe, *Fishing from the earliest times,* pp. 378-380.

12. Ch. L. Cutting, *Fish saving: a history of fish processing from ancient to modern times* (New York: Philosophical Library, 1956).

13. Aristotle, *Politics,* trans. by H. Rackham, The Loeb Classical Library (Cambridge, Mass.: Harvard University Press, 1932).

14. O. Bates, Ancient Egyptian Fishing, Harvard African Studies, I. Varia Africana (Cambridge, Mass.: African Department, Peabody Museum, Harvard University, 1917), pp. 199-272.

15. *Diodorus Siculus,* Book I, trans. by C. H. Oldfather, The Loeb Classical Library (Cambridge, Mass.: Harvard University Press, 1947).

16. L. Klebs, *Die Reliefs des Alten Reiches (2930-2475 BC) Material zur Aegyptischen Kulturgeschichte* (Heidelberg: Carl Winters Universitatsbuchhandlung, 1915).

17. P. Chimits, "Tilapia in Ancient Egypt," *FAO Fisheries Bulletin,* X:4 (October-December, 1957).

18. A. Lucas and J. R. Harris, *Ancient Egyptian materials and industries* (London, Edward Arnold, Ltd.), pp. 268-269.

19. W. Radcliffe, *Fishing from the earliest times,* p. 63.

20. Homer, The Odyssey, trans. by A. T. Murray, The Loeb Classical Library (Cambridge, Mass.: Harvard University Press, 1946), Book XII, pp. 251, 384.

21. Athanaeus, *The Deipnosophists,* Book XII, p. 521.

22. *Ibid.,* p. 119.

23. W. Radcliffe, *Fishing from the earliest times,* p. 224.

24. Arrian, *History of Alexander and Indica,* trans. by E. T. Robson, The Loeb Classical Library (Cambridge, Mass.: Harvard University Press), Book VIII, p. 28.

25. T. H. Corcoran, "Roman Fish Sauces," *Classical Journal,* 58:5 (February 1963), pp. 204–210.

26. Athanaeus, *The Deipnosophists,* Book II, p. 67.

27. B. Flower and E. Rosenbaum, *The Roman Cookery Book: A critical translation of cookery by Apicius* (London and New York: Peter Nevill Ltd., 1968).

28. Pliny, *Natural History,* trans. by H. Rackham, The Loeb Classical Library (Cambridge, Mass.: Harvard University Press, 1938), Book XXXI, p. 94.

29. F. J. Simoons, "Fish as forbidden food: The case of India," *Ecology of Food and Nutrition,* 3 (1974), pp. 185-201.

30. G. S. Robertson, *The Kafirs of the Hindu-Kush* (London: Lawrence and Bullen, Ltd., 1896), in Simoons, "Fish as forbidden food," *Ecology of Food and Nutrition,* p. 186.

31. R. C. F. Schomberg, *Kafirs and Glaciers: Travels in Chitral* (London: Martin Hopkinson, Ltd., 1938), in Simoons, "Fish as forbidden food," *Ecology of Food and Nutrition,* p. 186.

32. *Ibid.,* p. 190.

33. *Ibid.,* p. 185.

NOTES TO CHAPTER 2

1. Liebig's "Extractum Carnis," known in Germany, where it was manufactured, as "Liebig's Fleischextract," was a highly viscous, black, salty, treacle-like meat hydrolysate with a pronounced meaty flavor and taste. It was completely water soluble and was used in making soups and as a general condiment.

2. Report of the Commissioner, U.S. Commission of Fish and Fisheries, 1879, p. 139.

3. *Ibid.,* p. 141.

4. K. Bakken, "Technological Developments in Scandanavia," in *Fish in Nutrition,* ed. E. Heen and R. Kreuzer (London: Fishing News [Books] Ltd., 1962), pp. 419–424.

5. J. V. Baackes, "Targets connected with the manufacture of animal products.

Albumen and egg substitute industries," British Intelligence Objectives Sub-committee, 32, Bryanston Square, London W-1, August 1945, Bios Target Numbers C22/328, /331,/1672,/989,/335,/333,/334,/332,/327.

6. See Annotated Technical References.

7. The following percentage yields of fish protein concentrate were reported for various raw materials.
From ordinary fresh fish: 14
From dried cod: 50
From dried shrimp: 75
An average proximate percentage analysis of the product is reported as follows:
Protein (N x 6.25) 93.90 by weight of the finished product
Moisture 4.65
Salt 0.90
Fat Trace
Ash 0.80

8. K. Baaken, "Technological developments in Scandinavia," p. 419.

NOTES TO CHAPTER 3

1. A list of all agencies and institutions are included, with acronyms, at the beginning of the text. A chart of all participating agencies, delineating their position within the federal bureaucracy, is presented as exhibit 1.

2. Personal communication with Harry McKittrick, 1975.

3. Personal communication with Stewart Udall, 1975.

4. Personal communication with Harry McKittrick, 1975.

5. In March 1961, FAO sponsored an International Conference on Fish in Nutrition. The result of the meeting was an unpublished report strongly extolling the future of FPC as an aid to better nutrition.

6. Carr's interest in FPC later became so significant that, during a subsequent "Western Conservation Trip," the Undersecretary discussed the matter of FPC at length with President Kennedy on board Air Force One. Even later, in 1963, Carr's continuing enthusiasm led him to an audience with the Pope in Rome at which the potential of FPC and other marine protein resources was discussed. (Personal communication with Walter Pozen, 1975.)

7. The appendix contains detailed references to the various FPC production processes.

8. This company, whose primary product line involved health foods and special hospital feeding formulations, had been active since 1950 in developing a process to make FPC suitable for human consumption.

9. Personal communication with Ezra Levin, 1975.

10. Clause 3 indicated that "a food shall be deemed to be adulterated if it consists in whole or in part of any filthy, putrid or decomposed substance, or if it is otherwise unfit for food."

11. U.S. Department of Health, Education and Welfare, Food and Drug Administration, "Fish Protein Concentrate, Whole Fish Flour, Notice to Establish Definition and Standard of Identity," *Federal Register* 26, no. 178, 15 September 1961, p. 8641.

12. In September 1961, Congressmen Otis Pike (New York) and Hastings Keith (Massachusetts) did actually introduce identical bills in the House to amend Clause 3 of Section 402 (a) of the Food, Drug and Cosmetic Act to allow the use of FPC processed from whole fish. Both bills were killed in committee.

13. U.S., HEW, FDA, "FPC notice to Establish Definition and Standard of Identity," *Federal Register,* p. 8641.

14. Letter from the American Federation of Labor and Congress of Industrial Organizations (AFL-CIO) to the FDA, September 20, 1961.

15. Arnold H. Singal, "The Fish Flour Controversy–A Case Study of Government Regulation of Business," M.A. Thesis, Sloan School of Management, Massachusetts Institute of Technology, 1963, p. 59.

16. These were House bills 9101, 9102, 9331, and 10587 introduced during the 87th Cong., 2nd sess.

17. Letter from Stewart Udall to Detlev Bronk, President of the NAS-NRC, May 31, 1962.

18. Personal communication with Stewart Udall, 1975.

19. U.S., Department of Health, Education, and Welfare, Food and Drug Administration, "Fish; Definitions and Standards of Identity; Standards of Fill of Container. Fish Flour; Identity. Postponement of Public Hearing," *Federal Register* 27, no. 112, 9 June 1962, p. 5498.

20. Personal communications with Stewart Udall and Walter Pozen, 1975.

21. Hearings on the question of whether to modify Section 402 (a) (3) of the Food, Drug and Cosmetic Act to specifically permit the use of FPC were held before the Subcommittee on Health and Safety of the House Committee on Interstate and Foreign Commerce on August 8–9, 1962.

22. U.S., Congress, House, Committee on Interstate and Foreign Commerce, *Fish Protein Concentrate, Hearings before the Subcommittee on Health and Safety on H.R. 9101, H.R. 9102, H.R. 9331, H.R. 10587,* 87th Cong., 2nd sess., 1962.

23. The original NAS Committee on Marine Protein Resource Development

was composed of the following members: F. Daft (Chairman), J. B. Allison, W. M. Chapman, C. O. Chichester, J. Crutchfield, T. R. Donlan, P. Gyorgy, H. Hardy, J. Liston, H. S. Olcott, M. B. Schaefer, N. S. Scrimshaw, B. S. Schweigert, and L. Voris (NAS Staff Representative).

24. Personal communication with Stewart Udall, 1975.

25. General Foods Corporation was actually resuming work on experiments it had begun in the late 1940s and 1950s.

26. See chapter 6, "FPC in Canada."

27. The BCF believed that the FDA would find the idea of a schooling fish more acceptable since it would mean a homogeneous catch, and it would therefore be much easier to estimate average toxicity content.

28. Most recent work has focused on the less expensive, biological extraction process.

29. U.S., Congress, Senate, Committee on Commerce, *Hearing before the Subcommittee on Merchant Marine and Fisheries, a subcommittee of the Senate Committee on Commerce,* August 14, 1964, p. 2.

30. *Ibid.,* pp. 35-38.

31. *Ibid.,* p. 98.

32. Personal communication from Harold Crowther, 1975.

33. Personal communication from Walter Pozen, 1975.

34. The committee had held its second meeting on March 22-25, 1964, at which time it reviewed BCF research progress, international FPC development efforts, and the need for feasibility studies.

35. U.S., Congress, Senate, Committee on Commerce, *Fish Protein Concentrate, Hearings before the Senate Committee on Commerce on S 2720,* 89th Cong., 2nd sess., 1966, p. 40.

36. *Ibid.,* pp. 40-41.

37. In particular, two senators from large fishing states, Ernest Gruening of Alaska and Leverett Saltonstall of Massachusetts, as well as Senator Douglas of Illinois, made manifest their concern with the FDA position by inviting FDA Commissioner Larrick to a private luncheon on Capitol Hill where the entire menu was prepared using FPC. Since this type of experience had occurred once before for Commissioner Larrick, he refused to eat a bite, stating that he considered FPC "filthy and putrid." (Reported in the *Saturday Evening Post,* November 23, 1963.)

38. This shift in FDA policy, after nearly six years of negotiations on a food standard of identity for FPC, is significant. It is likely that the Agency stuck with the concept of the standard of identity for as long as possible because it offered the greatest degree of flexibility in the application of its authority

(the criteria for approval of a food standard are that the food be of acceptable quality, be clearly defined, and promote honest and fair dealing in the interest of the consumer). It is much easier for the FDA to reject a product under these nebulous terms than under the food additive regulations whose sole criterion is product safety.

39. Reference here is to the major food processors such as General Foods and excludes, of course, the well-known and continuing interest of the VioBin Corporation.

40. Speech by President Lyndon B. Johnson, February 10, 1966.

41. The USDI had withheld public disclosure of NAS Committee findings in the hope that, by avoiding the antagonism of the FDA, the regulatory agency might be persuaded to alter its position on the product. Since the FDA showed no sign of changing its policy and since pressure from the White House was increasing, the report was released by the BCF.

42. Food Additive Petition 6A 1991, notice of which appeared in the U.S., Department of Health and Welfare, Food and Drug Administration, Department of the Interior, "Notice of Filing of Petition for Food Additive Marine Protein Concentrate," *Federal Register* 31, no. 41, 2 March 1966, p. 3317.

43. Personal communications with Vincent Kleinfeld, Michael Markel, and Edward Byerley, 1975.

44. It is interesting to compare the difference in impact of the small but vocal congressional delegation from the fishing states with the overwhelming representation from the agricultural states.

45. Plans had called for the MMSU to be redesigned to provide the necessary continuous process data. However, the cost of the new instrumentation and equipment was so high that the project was severely cut back by the top management at the BCF. Eventually the equipment was ordered and did arrive, but the MMSU sat in its crates and was never even assembled due to lack of funds needed to install and operate the plant. (Personal communication with George Knobl, former director of College Park laboratory, 1975.)

46. A more recent example of inadequately developed technology is the recent introduction of the catalytic converter on automobiles, which may prove to create as much pollution as it eliminates. Yet another example is the core safety problems that have recently impeded the development of the much heralded nuclear power technology.

47. For the text of the bill see U.S., Senate Committee on Commerce, *1966 Hearings on FPC,* pp. 4–5.

48. It was reported in a personal interview with John Wedin, former counsel to the Senate Commerce Committee, that the wording for the FPC bill was actually taken directly from the Saline Water Bill.

49. U.S., Senate Committee on Commerce, *1966 Hearings on FPC,* pp. 18-19.

50. *Ibid.,* pp. 130-133.

51. Reported in *Food Chemical News,* May 23, 1966, p. 7.

52. Senate Report 1304 to accompany S. 2720, in U.S., Senate Committee on Commerce, *1966 Hearings on FPC.*

53. The rationale for the Council was that the line agencies concerned with marine affairs were to some extent duplicative, while letting certain problems fall "between the cracks." In some cases the agencies were even antagonistic. The role of the MSC, therefore, was to smooth out these problems and to recommend coordinated action.

54. Personal communication with Edward Wenk, Jr., former executive director of the MSC.

55. *Ibid.*

56. U.S., Congress, House, Committee on Merchant Marine and Fisheries, Hearings before a Subcommittee on Fisheries and Wildlife Conservation of the House Committee on Merchant Marine and Fisheries," 89th Cong., 2nd sess. 16-17 August 1966, p. 177.

57. The majority of the work completed by the BCF up to that point had focused on the processing of lean fish.

58. *Ibid.,* pp. 210-237.

59. *Ibid.,* pp. 220-223.

60. In a recent article in *Science,* FPC is used as an example of a federally funded demonstration project made necessary as the result of five kinds of uncertainty: (1) technological, (2) cost, (3) demand, (4) institutional, and (5) externality-related. The authors conclude that "The more appropriate government role is to reduce but not eliminate risks to private firms where significant public benefits are expected from the introduction of the new technology. A demonstration thus represents a bridging activity between the public and private sectors." W. S. Baer, L. L. Johnson, and E. W. Merrow, "Government-Sponsored Demonstration of New Technologies," *Science,* Vol. 196, No. 4293, 27 May 1977, pp. 952-955.

61. *Ibid.,* pp. 257-265.

62. *Ibid.,* pp. 265-268.

63. *Ibid.,* pp. 298-299.

64. U.S., Congress, House, *House Report 2165* to Accompany FPC Bill, 89th Cong., 2nd sess., 30 September 1966.

65. Personal communication with John Wedin, 1975.

66. Obviously, The BCF and VioBin had little choice but to accept the FDA

ruling, since they had applied for FDA approval. However, the point remains that the tolerance levels set by the FDA for foods used in the United States often bear little resemblance to the situation in the developing world. Despite this fact, the rulings of the U.S. FDA continue, in general, to be used as the world standard.

67. The lead agency concept was developed out of the realization that the MSC, as an analytic and coordinative body, could not run such a program. The idea was to place responsibility on the agency whose function most closely paralleled the task at hand.

68. Personal communication with Edward Wenk, Jr., 1975.

69. Edward Wenk, Jr., *The Politics of the Ocean* (Seattle: University of Washington Press, 1972), p. 315.

70. J. B. Cordaro, "A Review of AID's Past Efforts in Fish Protein Concentrate Development (1967-1970)," unpublished briefing paper for meeting of U.S. State Department Scientific and Technical Advisory Committee, Washington, D.C., June 18, 1971.

71. *Ibid.*

72. An agency memorandum recommended, in this regard, the formation of an Agency Committee for Unconventional Protein Program Development that would be prepared to recommend the development of other protein sources, such as oilseed, leaf protein, algal protein, protein from bacterial formation, or conventional food fortification with amino acids, should the FDA refuse to approve the manufacture of FPC. (Internal AID memorandum, December 20, 1966.) More recently the National Science Foundation has funded a comprehensive survey at MIT on foreseeable developments in the area of unconventional proteins.

73. U.S., Congress, House, Committee on Foreign Affairs, *Foreign Assistance Act of 1967, Hearings before the Committee on Foreign Affairs on H.R. 7099*, 90th Cong., 1st sess., 1967 (PL 90-147).

74. Marine Sciences Affairs, "Report to the President and Congress on Marine Resources and Engineering Developments," 1967.

75. The previous FDA commissioner, George Larrick, had retired a number of years earlier, thereby removing from the FDA a good deal of the opposition to FPC.

76. U.S., Department of Health, Education, and Welfare, Food and Drug Administration, "Food Additives Permitted in Food for Human Consumption. A Statement Concerning the Commercial Production of a Wholesome Protein Concentrate Made from Whole Fish for Human Consumption," *Federal Register* 32, no. 22, 2 February 1967, p. 1174.

77. Internal FDA memorandum, February 2, 1967.

78. FDA files on proceedings concerning FPC.

79. The stated rationale for withdrawal of the formal petition by the American Dry Milk Institute was that, in light of the pressing world need, hearings and judicial appeal would constitute an impediment to the U.S. foreign aid program. (From a personal communication with John Walsh, executive director of the American Dry Milk Institute, 1975.)

80. Personal communication with Harold Crowther, 1975.

81. Ocean Harvester, Inc., was a joint venture of SWECO, a California engineering and manufacturing corporation, and Star-Kist Foods, Inc., a wholly owned subsidiary of H. J. Heinz Co. The company thus combined engineering and manufacturing capability with competence in fish processing and food technology.

82. The Bureau had hoped to entice private sector involvement in the event of a cost overrun by offering the right of first access to the product.

83. Personal communication with Harold Crowther, former director BCF, 1975.

84. Wenk, *The Politics of the Ocean,* p. 314.

85. Those countries were: in *Latin America*—Chile, Brazil, Peru, Colombia, Panama, Nicaragua, El Salvador, Costa Rica, and Guatemala; in *Asia*—India, Pakistan, Ceylon, Thailand, the Phillipines, and South Korea; and in *Africa* —Morocco.

86. For example, there is some reason to believe the Korea was selected over Thailand or Indonesia for purely political reasons. (From a personal communication with Gerald Bernstein, IDMON, Inc., 1975.)

87. In the cooperative spirit of the PAS Agreement, George Parman, Director of the Food from the Sea Program, requested that the BCF provide AID with its own data concerning the capabilities of the 76 countries originally selected for inclusion in the survey. (Personal communication with George Parman, 1975.)

88. The Food from the Sea Service included the following staff: a director (food technologist), two division chiefs (one fisheries expert and one lawyer), a contract specialist, and a loan officer. J. B. Cordaro, a former member of that staff, has suggested "that the talents assembled were in most cases not suited for the tasks, that is, the qualifications lay in other fields," thus lending support to the notion, suggested earlier that AID was ill-equipped for its role as "lead agency." (J. B. Cordaro, "Fish Protein Concentrate: A Case Study of AID's Efforts to Introduce a New Protein Source into the Diets of People of Selected Developing Countries," unpublished manuscript, 1971.)

89. Interested firms in attendance included General Foods, W. R. Grace, National Biscuit, VanCamp Division of Ralston Purina, and H. J. Heinz.

90. See chapter 9 for a discussion of Astra's work with FPC.

91. Curiously, a violent explosion had devastated the New Bedford facility on June 1, 1967. The destruction from the blast was so extensive that major reconstruction had to take place before the plant could produce FPC under the AID contract.

92. Personal communication with Norman L. Brown, 1975.

93. See chapter 7 for details.

94. Message of President Johnson, quoted in U.S., Congress, House, Committee on Merchant Marine and Fisheries, *Hearings before the Subcommittee on Fisheries Wildlife Conservation of the House Committee on Merchant Marine and Fisheries on H.R. 4775, H.R. 7743, H.R. 15490, H.R. 15535,* 2nd sess., 1968, p. 88.

95. Statement of Harold E. Crowther in *Hearings before the House Subcommittee on Fisheries and Wildlife Conservation,* pp. 70–71.

96. Statement of Howard Wright in *Hearings before the House Subcommittee on Fisheries and Wildlife Conservation,* p. 114.

97. Statements of William Springer and Ezra Levin in *Hearings before the House Subcommittee on Fisheries and Wildlife Conservation,* pp. 9, 10, 41.

98. Statements of Edward Wenk and Howard Wright in *Hearings before the House Subcommittee on Fisheries and Wildlife Conservation,* p. 114.

99. Both Congressmen Dingell and Karth had expressed severe disapproval of the restrictive one-pound package rule of the FDA. Karth closely questioned an FDA official on the reasoning behind it, and Dingell instructed Subcommittee counsel Everett to "make some rather forceful inquiries of the FDA as to the reasons for the one-pound limitation, which appears to be lacking in any real or sufficient reason." (*Hearings before the House Subcommittee on Fisheries and Wildlife Conservation,* p. 118.)

100. *Ibid.,* p. 92.

101. *Ibid.,* p. 26.

102. The FDA had added the restriction on the use of the product by children less than eight years old in its original 1967 decision due to concern over the flurosis issue.

103. Ley had spoken informally on the FDA's position on food additives such as cyclamates and FPC. He commented that his experience in Japan had led him to be less concerned about the use of whole fish and that if another petition against the one-pound restriction were received it would be viewed sympathetically. (Memorandum for the Records, BCF, April 9, 1969.)

104. Letter from Dr. Herbert Ley to Dr. Frederick Seitz, NAS, May 26, 1969.

105. According to the terms of the original legislation for construction of the

EDP, formal certification by the FDA that FPC extracted with IPA met the terms of the Food, Drug and Cosmetic Act was required before work could begin on the program.

106. Internal BCF memorandum from N. Brown and K. Randolph to R. Finch, BCF, July 22, 1969.

107. *Ibid.*

108. J. B. Cordaro comments, "Looking back it appears that this was the first identifiable point where the program began to lose internal AID support," (that is, as reflected by the failure to fill the vacancies). (Cordaro, "Fish Protein Concentrate," p. 23.)

109. The FDA specification required that the PER value be *at least* equal to that of casein. The Alpine product apparently was at most .95 of the PER for casein as measured by standard, accepted statistical procedures. The VioBin Corporation subsequently disputed the validity of the procedures by which the PER had been calculated. (Personal communication with Ezra Levin, 1975.)

110. Cordaro has suggested that "for the fisherman it was a simple matter of economics: why should they fish for hake at 1 to 2 cents per pound when they can fish at 5 to 13 cents per pound. . . . It isn't in the fishermen's economic interest to fish for resources required for FPC at the price the producer can pay to make FPC a cheap protein ingredient." (Cordaro, "Fish Protein Concentrate," p. 30.)

111. The Chileans reportedly found the product unacceptable in odor, flavor, and color, as well as in protein content. For further details, see chapter 8.

112. Personal communication with George Parman and Ezra Levin, 1975.

113. Eventually this was to result in a $7 million cut in the BCF research budget. (Personal communication from Thomas Billy, NMFS, 1975.)

114. Personal communication with Thomas Billy, NMFS, 1975.

115. Recommendation No. 18, *Report of the White House Conference on Food, Nutrition and Health* (Washington, D.C.: U.S. Government Printing Office, December 1970).

116. U.S., Department of Health, Education, and Welfare, Food and Drug Administration, "Food Additives Permitted in Food for Human Consumption. Whole Fish Protein Concentrate," *Federal Register* 35, no. 150, 4 August 1970, p. 12390.

117. Consequently, a bill (S. 1273) was introduced in the Senate to extend the entire program for an additional fiscal year. The Senate passed S. 1273 in October 1971 on the recommendation of Senator Magnuson, but hearings were not held in the House. Thus, the attempt failed.

118. Personal communication from Thomas Billy, NMFS, 1975.

119. As a result, the following waste practices were in use: wash water was drained into the river; refrigerated brine water and fish and still waste were dumped into municipal sewers; and solid waste (such as, bones and scales) and sludge were disposed of in sanitary land fill.

120. For a complete analysis of the Chilean involvement with FPC see chapter 8.

121. Personal communication with J. B. Cordaro, 1975.

122. Internal AID memorandum for the files, 1971.

123. As of May 1972, the only new expenditure for FPC-related work was $100,000 allocated to Brazil to set up a bench-scale FPC demonstration unit.

124. In addition to its FPC advocacy role, the MSC had been responsible for a number of related ideas as well. For example, in March 1967 council director Wenk had become interested in the possibility of utilizing the prototype nuclear-powered cargo ship N.S. *Savanah,* (which was due to be mothballed because of exorbitant operating costs) as a floating FPC factory and demonstration ship. The *Savanah* was to be used (much as the S.S. *Hope* had been), to travel to less developed parts of the world to demonstrate the use and preparation of food from the ocean. An investigation of the plan was actually carried out through the Maritime Administration, and a report issued in April 1967 indicated that approximately $2 million would be needed to refit the vessel and another $3 million annually would be required for operation and maintenance. Although the costs made the project infeasible, it was put on the back burner with the thought that a floating FPC factory might someday cut the costs of production by eliminating the need to bring the fish ashore. Two years later, in August 1969, Marine Protein, Inc., of Panarama City, California, did again consider using the N.S. *Savanah* for the purpose of a floating FPC factory. The plan once again did not get off the drawing boards. (BCF internal memorandum.)

125. See article on risk and cost sharing between public and private sectors on high-risk projects: Baer, Johnson, and Merrow, "Demonstration of New Technologies, pp. 950–957.

126. H. M. Burgess, quoted in *Science,* July 31, 1971, p. 412.

127. Internal BCF files.

128. *Ibid.*

129. *Ibid.*

130. Personal communication from J. Philip Tice, Vice President for operations, World Food Systems, Inc. (formerly part of ARA Systems, Inc.), 1975.

131. Personal communication from Dr. Bruce R. Stillings, Nutrition Coordinator, Nabisco, Inc., 1975.

132. For further elaboration on the economics of FPC production see S. Tannenbaum, B. R. Stillings, and N. S. Scrimshaw, eds., *Economics, Marketing, and Technology of Fish Protein Concentrate* (Cambridge, Mass.: The MIT Press, 1974).

133. Report of the General Accounting Office to Congressman John D. Dingel on the Fish Protein Concentrate Program, May 25, 1973 (hereinafter cited as GAO Report, 1973), unpublished, p. 9.

134. According to the 1973 GAO Report the monthly percentage yield varied according to the different types of fish processed. Only 24 percent of the total FPC produced from hake and menhaden met the FDA specifications. An additional 11.9 tons of FPC were produced from anchovy which at that time was not approved by the FDA.

135. The conference was held on June 6–8, 1972, and produced the book by Tannenbaum, Stillings, and Scrimshaw, eds., *The Economics, Marketing, and Technology of Fish Protein Concentrate.*

136. The workers engaged at the EDP facility were private employees of the contractor, Ocean Harvesters, Inc., and therefore were not covered by civil service regulations.

137. U.S., Congress, House, Committee on Merchant Marine and Fisheries, *Hearings before the House Subcommittee on Fisheries and Wildlife Conservation of the House Committee on Merchant Marine and Fisheries,* 92nd Cong., 1st sess., 1972, p. 250.

138. *Ibid.,* p. 244.

139. The sites were evaluated on the basis of suitability, availability of facilities, and site and construction work needed. (Personal communication with Thomas Billy, NMFS, 1975.)

140. This was assuming that the estimates of fish necessary to complete the program remained constant and did not increase.

141. House Subcommittee on Fisheries and Wildlife Conservation, *1972 Hearings,* pp. 401, 405.

142. For example, after the NMFS estimate of 5200 tons of fish required to complete the program, an engineering firm, PEDCO, Inc., recommended that only 3600 tons were actually needed.

143. It was never fully established that an assured menhaden resource actually existed on the Gulf Coast. It is more likely that the fish supply would have been derived by diverting stocks from other uses such as fish meal.

144. U.S., House Subcommittee on Fisheries and Wildlife Conservation, *1972 Hearings,* p. 359.

145. *Ibid.,* p. 360.

146. Dingell also reportedly threatened to have a complete GAO investigation of the FDA undertaken if necessary.

147. *Ibid.*, p. 361.

148. *Ibid.*, p. 362.

149. GAO Report, 1973, p. 9.

150. House Subcommittee on Fisheries and Wildlife Conservation, *1972 Hearings*, pp. 326–331.

151. As a "lame duck," her power over appropriations was therefore effectively neutralized.

152. Personal communication with Thomas Billy, NMFS, 1975.

153. FDA files on the FPC petition.

154. U.S., HEW, FDA, "Food Additives Permitted in Food," *Federal Register*, p. 12390.

155. *Ibid.*, p. 9723.

156. U.S., Department of Health, Education, and Welfare, Food and Drug Administration, "Common or Usual Name for Plant Protein Products; Withdrawal of Textured Protein Products Proposal. Notice of Proposed Rule Making," *Federal Register* 39, no. 116, 14 June 1974, p. 205893.

157. GAO Report, 1973, p. 13.

158. The study urged the continued funding of research efforts as a "public good" in the likelihood that it would yield a commercially viable process. At the same time, it stated that the U.S. investment could not be justified solely on the basis of support for the fishing industry. (J. W. Devanney and G. Mahnken, *The Economics of Fish Protein Concentrate* [Cambridge, Mass.: MIT, Report No. MITSG 71-3, November 20, 1970].)

159. Personal communication with John Wedin, 1975.

160. The reference is to the U.S. Seafreeze Project, which is a federally funded program to construct two, 269-foot stern trawlers capable of competing with the Soviet and Japanese factory ships.

161. Harvey Sapolsky, *The Polaris System Development—Bureaucratic and Programmatic Success in Government* (Cambridge, Mass.: Harvard University Press, 1972), p. 241.

162. *Ibid.*, p. 244.

163. Factors cited in article by Baer, Johnson, and Merrow, "Demonstration of New Technologies," p. 956.

NOTES TO CHAPTER 4

1. In order to have a test case it would have been necessary to develop and

market the product, have it seized as adulterated, and then bring suit against the FDA. This would have required the cooperation of a private company, as well as the commitment of public funds to the development of a product likely to be found adulterated. Neither factor seemed likely and the weight of judicial precedent was against such an endeavor.

2. U.S., Department of Health, Education, and Welfare, Food and Drug Administration, "Fish; Definitions and Standards of Identity; Standards of Fill of Container. Fish Flour; Identity," *Federal Register* 27, no. 17, 25 January 1962, p. 740.

3. It is significant that the wording of the ruling did not state that FPC was filthy but that it would be *regarded* as filthy, thus implying a subjective, aesthetic judgment.

4. *Ibid.*

5. This policy also reflects the rough and rocky origins of the agency. The FDA had started its organizational life attempting to regulate the loosely controlled, filth-ridden food industry. Yet, even as late as 1961, manifestations of the "tough cop" image were still evident.

6. U.S., Congress, Senate, Committee on Commerce, *Fish Protein Concentrate, Hearings before the Senate,* Committee on Commerce on S. 2720, 89th Cong., 2nd sess., 1966, pp. 40–41.

7. Internal FDA memorandum, November 29, 1965.

8. Internal BCF memorandum, December 6, 1965.

9. Internal BCF memorandum of meeting, December 9, 1965.

10. According to an internal FDA memorandum, the FDA believed that "originally the proponents of whole FPC did not intend domestic use, but mainly desired a cloak of respectability for the fish flour. The Food Additive regulation will serve that purpose." (Internal BCF memorandum of meeting, December 9, 1965.)

11. Personal communication with Vincent Kleinfeld, 1975.

12. On February 28, 1966, a letter of this nature was sent by Senator Warren Magnuson, chairman of the Senate Committee on Commerce, to Secretary of Health, Education, and Welfare John W. Gardner urging "early appraisal and approval of FPC as recommended by the NAS Committee." (Internal BCF files.)

13. Personal communication with Vincent Kleinfeld, 1975.

14. BCF transmitted six lead assays, three of which contained levels below 3 ppm, while the others indicated 5, 6, and 7 ppm respectively. (Internal BCF memorandum.)

15. Minutes of the FDA-BCF meeting of April 28, 1966, p. 4.

16. The report by Dr. Zipkin on the effect of fluoride on teeth and bones indicated that levels of fluoride in the diet in concentrations of greater than 10 ppm would probably cause *cosmetic mottling* of teeth, and the FPC could not therefore be recommended for use in children's diets in the United States.

17. Letter from Donald McKernan to Robert Roe of the FDA, May 5, 1966.

18. Chapman pointed out that at the same time the FDA was considering the acceptability of FPC, a surplus of nonfat dry milk (NFDM) was being purchased by the Department of Agriculture. This purchase was of vital importance to the dairy industry because it needed to dispose of the protein part of the milk (which competed for the same market as FPC) in order to produce the butterfat for which there was a separate and lucrative commercial market. The FDA ruling on FPC also prohibited the use of fish oils, which had been used safely in Europe for some time as a component of margarine. However, the European dairy industry is not subsidized to the same extent as the dairy industry in the United States; thus, the prohibitions on the use of fish oil, plus the resistance to FPC directly, Chapman argued, may have been directed toward protecting the dairy industry from competitive products. (Personal communication from Wilbert Chapman, November 4, 1966.)

19. U.S., Senate Committee on Commerce, *FPC Hearings,* 1966, pp. 18-19; and U.S., House of Representatives, Committee on Merchant Marine and Fisheries, Subcommittee on Fisheries and Wildlife Conservation, *Hearings,* 89th Cong. 2nd sess., August 16-17, 1966, p. 197.

20. An official of the NMFS, Thomas Billy, has estimated that the Bureau spent roughly $5 million out of a total of $14 million in gaining FDA approval.

21. Informal meeting held at the BCF with E. Gordon Robbins and John Bausch, lawyers for Armour and Company.

22. The most effective arguments that had been raised related to the deception of the consumer and to unfair competitive practices. Both arguments would be negated by special packaging regulations.

23. Internal BCF memorandum for the files of a meeting between J. Slavin and D. Allen at the BCF and attorney M. Markel, December 7, 1966.

24. U.S., Department of Health, Education, and Welfare, Food and Drug Administration, "Food Additives Permitted in Food for Human Consumption, A Statement Concerning the Commercial Production of a Wholesome Protein Concentrate Made from Whole Fish for Human Consumption," *Federal Register* 32. no. 22, 2 February 1967, p. 1174.

25. Internal FDA memorandum, February 2, 1967.

26. Internal FDA memorandum by William Goodrich, FDA staff counsel.

27. Edward Wenk, Jr., *The Politics of the Ocean* (Seattle: University of Washington Press, 1972), p. 554, footnote 26.

28. Personal communication with John Walsh, executive director of the American Dry Milk Institute, 1975.

29. A Petition by the Swedish Astra Group in January 1968 to make FPC from whole herring was not approved due to a fluoride content of 180 ppm.

30. Letter from Dr. Nevin Scrimshaw to Harold Crowther, director of the BCF, June 13, 1968.

31. Internal BCF memorandum, July 30, 1968.

32. Alpine also discontinued its efforts because the contract with AID had failed and was subsequently terminated (discussed later in detail). This resulted in a change of management at Alpine and little further interest.

33. Internal BCF memoranda prepared by Dr. George Knobl, director of the College Park Laboratory, August 5 and 11, 1969.

34. On January 29, 1970, the VioBin Corporation also petitioned the FDA and added the threadherring to this list.

35. Letter from Louis Buckley, head of the Division of Regulations and Petitions Control to the director of the BCF, April 1, 1970.

36. At virtually the same time that NAS was downplaying the FDA's concerns, a letter from the Director of the American Dry Milk Institute urged that the FDA should rather "completely review the existing whole fish protein regulations . . . instead of embarking precipitously upon an expansion of that program." (Internal BCF memorandum, May 10, 1970.)

37. Letter from the NAS Subcommittee on Marine Protein Resources to Leo Friedman of the FDA, May 18, 1970.

38. U. S., Department of Health, Education, and Welfare, Food and Drug Administration, "Food Additives Permitted in Food for Human Consumption. Whole Fish Protein Concentrate," *Federal Register* 35, no. 150, 4 August 1970, p. 9723.

39. Supporters of Ezra Levin and the VioBin Corporation have suggested that personal animosity may have also played a role in the rejection of the VioBin process.

40. U.S., Department of Health, Education, and Welfare, Food and Drug Administration, "Alpine Geophysical Associates, Inc. Notice of Filing of Petition for Food Additives," *Federal Register,* no. 33, 18 February 1971, p. 3145.

41. Letter from Louis Buckley, Petitions Control Branch of the FDA, to Michael Markel, attorney for Alpine Geophysical Associates.

42. On March 30, 1972, the FDA published a proposal in the *Federal Register* creating voluntary nutrition labeling regulations. Because protein content was to be expressed to the nearest gram, fishery products were adversely affected,

due to seasonal variations in protein content. To comply with the proposed regulations, expensive analysis and frequent label changes would also be required.

43. U.S., HEW, FDA, "Food Additives Permitted in Food," *Federal Register,* p. 9723.

44. *Ibid.*

45. *Ibid.*

46. Arnold H. Singal, *The Fish Flour Controversy—A Case Study of Government Regulation of Business* (M.A. Thesis, Sloan School of Management, Massachusetts Institute of Technology, 1963), pp. 66–76.

47. Except in cases of "clear and present" danger due to toxicity.

48. Singal, *The Fish Flour Controversy,* pp. 66–76.

NOTES TO CHAPTER 5

We wish to express our debt to Arnold H. Singal, whose well-written and researched Master's thesis, "The Fish Flour Controversy—A Case Study of Government Regulation of Business" (M.A. thesis, Sloan School of Management, MIT, 1963) provided the basis for this chapter. With the permission of Mr. Singal we have paraphrased and quoted directly from his thesis throughout this chapter.

1. Federal Food, Drug, and Cosmetic Act as amended, 21 U.S.C. S342(a)(3), 1938.

2. 32 *Federal Register* 1174.

3. Introduction by Hamilton, in S. Wilson, *Food and Drug Regulation,* 1942.

4. *United States* v. *62 Packages Marmola Prescription Tablets;* 320 U.S. 43 (1943).

5. *United States* v. *Dotterweich;* 320 U.S. 282 (1943).

6. Regier, *Struggle for Federal Food and Drug Legislation,* 1 LAW & CONTEMP. PROB. 3 (1933).

7. Hart, *Adulteration of Food Before 1906,* 7 FOOD DRUG COSM. L.J. 5 (1952).

8. *Ibid.*

9. Vincent Kleinfeld, *Legislative History of the Food, Drug, and Cosmetic Act,* 1 FOOD DRUG COSM. L.Q. 532 (1946).

10. Dunn, *Introduction,* 1 FOOD DRUG COSM. L.Q. 4 (1946).

11. Hart, *Adulteration of Food,* p. 5.

12. Robbins, *The Food Industry,* 5 FOOD DRUG COSM. L.J. 297 (1950).

13. Hart, *Adulteration of Food,* p. 5.

14. Regier, *Struggle for Federal Food and Drug Legislation,* p. 3.

15. C. Wilcox, *Public Policies Toward Business,* Homewood, Illinois: R. D. Irwin, Inc., 1960, p. 201.

16. Regier, *Struggle for Federal Food and Drug Legislation,* p. 3.

17. Wilcox, *Public Policies Toward Business*, p. 210.

18. S. 1944, 73rd Cong., 1st sess., 1933.

19. Public Law No. 447, U.S. Stat. L. Volume 52, 1938.

20. Congressional Record, Volume 83, 1938, pp. 11543–11549.

21. U.S. Quartermasters Food and Container Institute, "Study of the Bases for Changing Food Attitudes," Washington, D.C.: U.S. Government Printing Office, 1958.

22. Kleinfeld, Objection to Federal Regulation Notice of January 25, 1962, filed with the FDA, 1962.

23. B. L. Oser, "The Wholesomeness of Fish Protein Concentrate with Particular Reference to the Presence of 'Extraneous Matter,'" (produced on Contract with the Bureau of Commercial Fisheries, USDI), Food and Drug Research Laboratories, Inc., November 19, 1962, p. 3.

24. S. A. Goldblith and E. L. Wick, "Analysis of Human Fecal Components and Study of Methods for Their Recovery in Space Systems," (conducted at M.I.T. on contract with the U.S. Air Force), Contract No. AF 33 (616)-6136, 1961.

25. B. L. Oser, *The Wholesomeness of Fish Protein Concentrate,* p. 6.

26. *Ibid.,* pp. 6–7.

27. Personal communication from Leo Friedman.

28. B. L. Oser, *The Wholesomeness of Fish Protein Concentrate,* p. 7.

29. *Ibid.,* p. 8.

30. *Ibid.,* pp. 9–10.

31. *Ibid.*

32. *Ibid.,* p. 11.

33. *United States* v. *St. Louis Coffee and Spice Mills,* 189 Fed. 191 (E.D. Missouri 1909).

34. *United States* v. *Sprague,* 208 Fed. 419 (E.D. New York 1913).

35. 1 CCH FDC Law Reporter. S. 3027 (2nd ed.).

36. *United States* v. *44 Cases, Vivano Spaghetti with Cheese,* 101 F. Supp. 658 (E.D. Illinois 1951); *United States* v. *Lazers,* 56 F. Supp. 730 (N.D. Iowa 1944); *United States* v. *Swift,* 53 R. Supp. 1018 (N.D. Georgia 1943).

37. No. 8917 Notices of Judgment (D. Maryland 1942).

38. No. 7808 Notices of Judgment (W.D. Michigan 1948).

39. Underwood, *Keeping Food Clean,* 6 Food Drug Cosmetic L.J. 303 (1951).

40. Federal Food, Drug and Cosmetic Act as Amended, 21 U.S.C. S 336 (1938).

41. Adelman, 402 (a) (3) and the de Minimis Problem, 8 FOOD DRUG COSM L.J. 49 (1953).

42. *United States* v. *588 Cases of Iona Tomatoes,* reported in 2 CCH F.D.C. Law Reporter 7260 (D.D. Illinois 1935).

43. *62 Cases of Jam* v. *United States,* 340 U.S. 593 (1951).

44. Personal communication with Vincent Kleinfeld, 1975.

45. Federal Food, Drug and Cosmetic Act as Amended, 21 U.S.C. S 375 (a) (1938).

46. Federal Food, Drug and Cosmetic Act as Amended, 21 U.S.C. S 333 (1938).

47. *United States* v. *Roma Macaroni Factory,* 75 F. Supp. 663 (N.D. California 1947).

48. See for example, Max Milner, "New Methods to Detect and Eliminate Insect-Infested Grain," *Advances in Food Research,* Volume VIII (New York: Academic Press, Inc., 1958).

49. Statement of Dr. Herbert Ley, before the House Committee on Science and Astronautics, December 1, 1969

50. Stephen Hull McNamara, "Some Legal Aspects of Providing New Foods for Hungry Populations," *Food Product Development,* March 1975, p. 57.

51. *Ibid.,* p. 59.

52. *Ibid.*

NOTES TO CHAPTER 6

1. Conference on Fish Protein Concentrate, Ottawa, Canada, October 24 and 25, 1967, *Canadian Fisheries Reports,* No. 10 (July 1968).

2. For further description of this product see chapter 2 and appendix 1.

3. D. R. Idler, "The development and scope of the Halifax process," paper presented to the Conference on Fish Protein Concentrate, Ottawa, *Canadian Fisheries Reports,* pp. 45–56.

4. *Ibid.,* p. 45.

5. N. Dambergs, "Acetone water mixtures for the extraction and rapid estimation of fats of biological materials, particularly fish products." *Journal of the Fisheries Research Board of Canada,* 13:(6) (1956), pp. 791–797; "Extractives of fish muscle. 2. Solvent-water ratio in extraction of fat and water

solubles," *Journal of the Fisheries Research Board of Canada,* 16:1 (1959), pp. 63-71; "Isopropanol-water mixtures for the production of fish protein concentrate from Atlantic herring (Clupea Harengus)," *Journal of the Fisheries Research Board of Canada,* 26:7 (1969), pp. 1919-1923; "Isopropanol-water azeotrope as solvent in the preparation of fish protein concentrate from herring (Clupea Harengus)," *Journal of the Fisheries Research Board of Canada,* 26:7 (1969), pp. 1923-1926.

6. Department of Industry, Foods Products Branch, Ottawa, 1964, Catalogue Number ID 3-964.

7. B. Weinberg, "The current status of fish protein concentrate," paper presented to the Conference on Marine Sciences, University of Prince Edward Island, Charlottetown, P.E.I., August 16-18, 1970.

8. Conference on Fish Protein Concentrate, Ottawa, *Canadian Fisheries Reports.*

9. Perhaps because the FDD did not fully agree with U.S. FDA's stand on this matter, and the political pressures in Canada were less severe than in the United States.

10. B. Weinberg, "The current status of fish protein concentrate," p. 11.

11. Canadian Foou and Drug regulations for fish protein (fish protein concentrate), Schedule of Amendments No. 130, passed by Governor-in-Council under P.C. 1970-1619 of September 16, 1970, and published in *The Canada Gazette,* October 14, 1970.

12. B. Weinberg, "The current status of fish protein concentrate," p. 12.

13. *Oceanology,* 4:1 (Thursday, January 4, 1968).

14. *Ibid.*

15. *Chemical Week,* August 3, 1968.

16. The Chronicle Herald, Halifax, N.S., April 11, 1969, and letters from W. J. Lavigne for the Minister of Fisheries and Rural Development, February 17, 1969.

17. It is believed that H. W. Duval had at the time invested very little of his own money in the venture.

18. B. Weinberg, "The current status of fish protein concentrate," p. 12.

19. In January, Duval made euphoric predictions about building other plants in Canada and elsewhere in the world.

20. The support given by the U.S. Government to the FPC project in the United States was comparable as far as assistance to the technical and scientific effort was concerned; however, unlike the Canadian FDD, the U.S. FDA appeared for a long time to do everything in its power to prevent the development of FPC.

NOTES TO CHAPTER 7

1. Protein Advisory Group of the United Nations.

2. "Fishery Country Profile. Food and Agriculture Organization of the United Nations," FAO Document FID/CP/MOR Rev. 1, March 1975. Morocco. FAO reports a production of 208,000 tons of fish "for animal food and other purposes." There is no indication, however, that any significant fraction of this amount went into products other than fish meal. An AID survey team reported that in 1966 more than 75 percent of the sardine catch went to fish meal manufacture (with a very small amount of this used to make FPC by solvent extraction), about 20 percent was canned, and the remainder marketed fresh, frozen, or salted. See also *Africa Report–War on Hunger–Food from the Sea,* Report by survey team, USAID (Leroy Christy, AID/War on Hunger/Food from the Sea Program; Benjamin Jones, Bureau of Commercial Fisheries (BCF); John A. Dassow, BCF; Salvatore DiPalma, BCF).

3. As estimated 55 percent of total sardine landings went to fish meal at that time. The remainder was canned, primarily for export.

4. John J. Blake reports a maximum prediction rate of 500 kilograms per day. (Personal communication, 1970.)

5. Abdellah Jaidi, "Le concentré de protéines," 1er Congrès International Industries Alimentaires et Agricoles, Abidjan, Côte d'Ivoire, 14–19 Decèmbre 1964.

6. "World-Wide Developments on Fish Flour," Nutrition Document R.8/Add. 10, July 8, 1960, WHO/FAO/UNICEF-PAG, August 1960 Meeting, Washington, D.C.

7. Authors' translation. See Jaidi, "Le concentré de protéines."

8. G. D. Kapsiotis, "Manufacture of fish protein concentrate from sardines at the factory of Azote-Union at Safi," Report No. 14489, September 1961, FAO.

9. Jaidi, "Le concentré de protéines."

10. See Jaidi, "Le concentré de protéines"; Lynn Catoe, personal communication, 1975. (Ms. Catoe was with the Catholic Relief Services and the Ministry of Public Health from 1955 to 1962.)

See also G. D. Kapsiotis, "The potential utilization of fish protein concentrates with special reference to the Moroccan FPC," PAG Document R.8/Add. No. 19, PAG (FAO/WHO/UNICEF), July 1965 Meeting, Rome. This report gives several examples of formulations suggested by Dr. François Rémy that were tried in Morocco. For example:

Fish Rice (6 persons)
350 g rice
50 g oil or margarine
3/4 1 water or meat broth
salt, pepper
70 g FPC

FPC lentils (6 persons)
250 g rice
50 g oil or margarine
2 big green peppers
2 big tomatoes
1 onion
90 g FPC

Kesra (Bread) with FPC
For traditional preparation of one Kesra:
500 g wheat flour
yeast, salt, ginger, aniseed
40 g FPC

Biscuits (homemade)
1,000 g wheat flour
5 g honey
300 g sugar
5 g yeast
10 g aniseed
100 g margarine
75 g FPC
(The biscuits are of 10 g size.)

11. Helene Kadane, Consultative Council of Jewish Organizations, personal communication, 1975. At the same time that these early acceptability studies were being conducted with FPC, a shipload of MPF (Multi-Purpose Food), supplied by Meals for Millions Foundation, arrived in Morocco. Consequently, many of the same people involved in the FPC testing found themselves conducting similar tests with MPF at the same time. (Catoe, personal communication, 1975.)

12. "Fish Protein Concentrate (FPC) Market Feasibility Study," Non Capital Project Paper (PROP), USAID Project No. 931-11-590-866, PROP Amendment: No. 1, July 14, 1970 (hereinafter cited as "FPC Market Feasibility Study").

13. "Société Nationale de Farine Alimentaire de Poisson—'SONAFAP,'" Report of the Bureau d'Etudes et de Participations Industrielles, Rabat, Morocco, July 1966 (hereinafter cited as "SONAFAP Report"), p. 5.

14. "Exploratory Mission for Fish Protein Concentrate Plant in Morocco," Report to UNIDO/FAO, New York-Rome, June 12, 1967, p. 7.

15. Blake, personal communication, 1970.

16. John J. Blake, "Production of fish-protein concentrate from Moroccan sardines," in *Production of Fish Protein Concentrate,* Report and proceedings of the Joint UNIDO/FAO Expert Group Meeting, Rabat, Morocco, December 8-12, 1969, *Part III. Proceedings of the Meeting* (New York: United Nations, 1972), pp. 55-74.

17. "Exploratory Mission for FPC Plant in Morocco," p. 7.

18. *Ibid.*, p. 3.

19. *Ibid.*, pp. 8–11.

20. See chapter 3, "FPC in the United States," for a discussion of the Food from the Sea Program.

21. Arturo Chavez, personal communication, 1975. (Mr. Chavez is Senior Technical Advisor, Technical Advisory Division, United Nations Development Program [UNDP], New York.)

22. U.S. Department of State, Telegram, Unclassified, State 123991, July 25, 1969.

23. Norman L. Brown and Harry Miller, Jr., "Experimental Production of fish protein concentrate (FPC) from Mediterranean sardines," *Commercial Fisheries Review,* 31:10 (1969), p. 30.

24. *Africa Report—War on Hunger—Food From the Sea,* Report by survey team, USAID.

25. *Ibid.*

26. Kapsiotis, "The potential utilization of FPC with reference to Moroccan FPC," pp. 2, 3.

27. Catoe, personal communication, 1975.

28. G. D. Kapsiotis, "Possibilities for development of protein-rich food mixtures for infants and children in Morocco, Algeria and Tunisia," FAO Restricted Report, FAO, Rome, 1965.

29. Catoe, personal communication, 1975; and J. B. Cordaro, personal communication, 1975. (Mr. Cordaro was on the staff of the Office of the War on Hunger, Food from the Sea Program, AID.)

30. Philip Birnbaum, Personal communication, 1975. (Dr. Birnbaum was assistant director and director of the AID mission in Rabat in the years 1967–1969.)

31. Dr. John H. Blake, who had been a member of the 1967 UNIDO/FAO Exploratory Mission, had been sent back to Agadir by UNIDO to obtain the engineering data needed for redesign of the SONAFAP plant.

32. *Production of Fish Protein Concentrate,* Report and Proceedings of the Joint UNIDO/FAO Expert Group Meeting, Rabat, Morocco, 8–12 December 1969 (New York: United Nations, 1972).

33. *Ibid.*

34. J. Blake, "Production of fish-protein concentrate from Moroccan sardines."

35. J. B. Cordaro, "The USAID program for evaluating and promoting FPC," in *Production of Fish Protein Concentrate,* Report and proceedings of Joint UNIDO/FAO Expert Group Meeting, Rabat, Vol. II, pp. 155–162.

36. U.S. Department of State, Telegram, Unclassified, State 123991, July 25, 1969.

37. J. W. Devanney, and G. Mahnken, "Economic aspects of ocean activities. the economics of fish protein concentrate," Report No. MITSG 71-3, Massachusetts Institute of Technology, Cambridge, Mass., USA, 1970.

38. *Production of Fish Protein Concentrate,* Report and proceedings of Joint UNIDO/FAO Expert Group Meeting, Rabat, 8-12 December 1968.

39. *Ibid.*

40. In a letter dated June 7, 1968, Mr. M. Mautner, Chief of the Food and Light Industries Section of UNIDO's Industrial Technology Division, wrote to Paul-Marc Henry, Assistant Administrator and Associate Director of UNDP's Bureau of Operations and Planning, that he anticipated "additional costs" of $1.5 to $2 million for the future commercial plant with capacity between 5,000 and 10,000 tons per year of FPC.

41. "Final Report. Fish Protein Concentrate Plant: Feasibility Study (DP/MOR/70/533)," UNIDO, Vienna, Austria, Document No. UNIDO/TCD/SF. May 21, 1974, p. 4. The report that was made available for this study was in French. All quotations in English are translations by the authors.

42. *Ibid.,* pp. 5-6.

43. AID/Washington was notified by cablegram that the Mission in Rabat had received a letter from the government requesting cancellation of the proposed market study. (Lynn Catoe and A. Gordon MacArthur, personal communication, 1975.) The actual date of cancellation seems to have been either April 25, 1972 (Joel Bernstein, personal communication, 1975) or June 29, 1972 (USAID Telegram, Unclassified, Rabat 4836, October 9, 1975).

44. "Final Study: FPC Plant," p. 11.

45. Catoe, personal communication, 1975; A. Gordon MacArthur, Morocco Desk Officer, AID, from 1969 to 1971. (Since 1971 Program Officer at USAID/Rabat), personal communication, 1975; Mohamed Benjelloun, Counselor and Charge d'Affaires ad interim, Embassy of Morocco, Washington, D.C., personal communication, 1975.

46. UNDP, Quarterly Report, July 1 to September 30, 1971, submitted by UN Resident Representative, Rabat, to UNDP, New York.

47. UNDP, Memorandum to the Files, Resident Representative, Rabat, February 21, 1972.

48. UNDP, Letter to AID, February 28, 1972.

49. UNDP, Note to File, reporting on a meeting between Arturo Chavez (UNDP) and J. B. Cordaro (USAID), March 1972 (date uncertain).

50. Gerdt Lövold, Letter to FAO, May 25, 1972.

51. UNDP, Memorandum to the Files, UN Resident Representative, Rabat, June 20, 1972.

52. UNDP, Letter from UN Resident Representative, Rabat, to UNDP, New York, June 29, 1972.

53. USAID Telegram Unclassified Rabat 4836, October 9, 1975.

54. UNDP, Letter from UN Resident Representative, Rabat, to UNDP, New York, June 29, 1972.

55. USAID Cablegram, Unclassified, Rabat 4836, October 9, 1975.

56. Joel Bernstein, personal communication, 1975.

57. Elmer J. Moore, 1970 Memorandum to Samuel Butterfield, Associate Assistant Administrator, AID, Bureau for Technical Assistance, dated February 5, 1970, Subject: Morocco FPC Market Study.

58. "FPC Market Feasibility Study."

59. Irwin Hornstein, Deputy Director, Office of Nutrition, Bureau for Technical Assistance, AID, personal communication, 1975.

60. "FPC Market Feasibility Study," p. 2.

61. Mautner, letter, June 7, 1968.

62. J. B. Cordaro, Memorandum to the files, July 22, 1970, Office of Agriculture and Fisheries, Bureau of Technical Assistance, AID.

63. Letter from UN Representative, Rabat, to UNDP, New York, June 29, 1972.

NOTES TO CHAPTER 8

1. For a detailed discussion of this program see Peter Hakim and Giorgio Solimano, "Supplemental Feeding as a Nutritional Intervention: The Chilean Experience in the Distribution of Milk", International Nutrition Planning Program Discussion Papers 75/2 (Cambridge, Mass.: Center for International Studies, MIT, 1975).

2. "Notes on Fish Flour for Human Consumption," prepared by the Nutrition Division of FAO for consideration of the Protein Study Group at its meeting on March 19 and 20, 1956, at United Nations Headquarters, New York, WHO/FAO/UNICEF Nutrition Panel R. 8, March 15, 1956, p. 11.

3. A. Vergara, "Extensive Report on the Fish Flour Experiment in Chile," FAO Document FAO/54/3/1652, p. 8.

4. P. Hakim and G. Solimano, "Supplemental Feeding as a Nutritional Intervention."

5. Agreement Concerning a Plan of Operations for a Fish Flour Plant at Quin-

tero, Chile, February 16, 1956 (translated from Spanish by UN translation section using draft English translation), p. 1. (Author's emphasis.)

6. Layton E. Allen, personal correspondence, November 26, 1974.

7. *Ibid.*

8. Layton E. Allen, "Fish Flour Production in Chile," *Fishing News International,* 2:1 (January-March 1963).

9. Allen, personal correspondence, November 26, 1974.

10. Julio Santa Maria, personal correspondence, June 25, 1975. (Author's translation.)

11. Contract Regarding the Production of Fish Flour for Human Consumption, Contract between Government and ISESA, signed February 16, 1956 (UNICEF document, translated from Spanish by UN translation section).

12. UNICEF correspondence, April 3, 1958.

13. Memorandum on the start up of operations in the Fish Meal Plant for Human Consumption, (Quintero—March-April 1958), Memorandum no. 161, Nutrition Sub-Department, National Health Service, Santiago, Chile, April 21, 1958, p. 6.

14. Erling K. Odland, "Report Quintero Fish Flour Plant—test runs made during June-July 1961," UNICEF internal report, unpublished, August 18, 1961, p. 1.

15. *Ibid.*, p. 3.

16. Acta de la Sesion No. 556, Consejo Nacional de Salud, December 13, 1961, p. 3.

17. See chapter 6 on another private enterprise, the Astra Group of Sweden, which possessed all of these assets and which would go about FPC development in a far different manner. Perhaps the most important element possessed by Astra and not by ISESA was familiarity with long-term product development.

18. Acta de las Sesion No. 556, December 13, 1961, p. 5.

19. Hakim and Solimano, "Supplemental Feeding as a Nutritional Intervention," p. 11.

20. Acta de la Sesion No. 556, December 13, 1961, p. 5.

21. Charles F. Lee, Report to Pan American Health Organization on Quintero Fish Flour Plant (Washington, D.C., 1965), p. 1.

22. *Ibid.*

23. An estimate derived from all documents relating to Quintero available to the authors.

24. Santa Maria, personal correspondence, June 25, 1975.

25. *Ibid.*

26. Lee, Report to PAHO, p. 6.

27. "Estudio de la Aceptabilidad de un Alimento Proteico Para Preescolares A Largo Plazo en La Provincia de Curico," Servicio Nacional de Salud, 1969.

28. Peter Hakim, personal communication, 1975.

29. *Ibid.*

30. World Bank Report on Possibilities of Technical Assistance Loan to Chile, Annex IV: Nutrition and Fisheries, Washington, D.C., 1973.

31. Project Summary of "Industrial Protein Complex" by CPI (IFOP-CORFO), 1973, p. 1.

32. Max Rutman, personal communication, 1975.

33. World Bank Report, Annex IV, 1973, p. 16.

34. *Ibid.*, p. 16.

35. Peter Hakim, unpublished document, 1973.

36. Hakim, personal communication, 1975.

37. World Bank Report, Annex IV, 1973, p. 15.

38. Hakim, unpublished document, 1973.

39. Giorgio Solimano, personal communication, 1975.

40. F. Monckeberg, E. Yanez, D. Ballester, C. O. Chickester, and Jung-Ching Lee, "Effects of Processing on the Nutritive Value of WSB-Skim Milk Mixtures for Infants," Study supported by USAID grant no. csd-3646, conducted by the Department of Nutrition, University of Chile, and the Department of Food and Resource Chemistry, University of Rhode Island, November 1974, p. 1.

41. Report of AID Survey Conducted in Latin America, Food from the Sea Service, Office of War on Hunger, Washington, D.C., September 1967.

42. *Ibid.*

43. *Ibid.*

44. Santa Maria, personal correspondence, June 25, 1975.

45. *Ibid.*

46. Memorandum No. 338, Ministerio de Salud Publica, Programas Alimentarios, December 1969, p. 9 (author's translation).

47. "Commercial Feasibility of Fish Protein Concentrate in Developing Countries," Report prepared for USAID by General Oceanology, Inc., Cambridge, Massachusetts, 1970, p. 5.

48. *Ibid.*, p. 4.

NOTES TO CHAPTER 9

1. The Astra endeavor is called a success because this company has sold approximately 30,000 metric tons of FPC since it began commercial production in 1969. It must be understood, however, that not all of this FPC is for human consumption.

2. "Fish Protein Concentrate," publication of Astra Nutrition, Mölndal, Sweden, 1967, p. 9.

3. Reference is made in various Astra documents to these technical difficulties, but no details are given. While Astra public relations publications are extensive, specific details about the course of its FPC development are of a proprietary nature and are not available.

4. H. A. Arnold, "Report of Visit to Astra Nutrition," document of National Council on Marine Resources and Engineering Development, June 22, 1967, unpublished memo, Washington, D.C.

5. "Prot-Animal" is not fish meal; it is a solvent extracted product with high protein and low fat contents.

6. See chapter 3.

7. Arnold, "Report of Visit to Astra Nutrition."

8. *Ibid.*

9. Department of State Airgram from American Embassy, Stockholm, November 26, 1968.

10. Quoted in "Swedish firm plans to produce FPC in converted whale factory ship," *Fishing News International,* November 1969.

11. See chapter 6.

12. Astra Annual Report, Sodertalje, Sweden, 1973, pp. 2, 14.

13. Lloyd's Weekly Casualty Reports, April 14, 1974.

14. Astra Annual Report, Sodertalje, Sweden, 1974, p. 12.

15. Astra Protein Letter, 1:2 (September 1973), p. 3.

16. J. S. Annan, Director, Division of External Relations and General Services, World Food Program, personal correspondence, June 3, 1975.

17. This quantity must be considered in contrast to a total of about one million tons of various food products received by the WFP in a year.

18. Gosta Oscursson, Lars-Borje Sjoberg, Ole M. Rostad, "Fish Protein Concentrate—the Protein Upgrader" (Astra publication, Mölndal, Sweden), 1975, p. 5.

19. *Astra Protein Letter,* 4:1 (January 1976), p. 5.

20. Rostad, personal communication, 1976.

21. Astra estimates the price per pound of protein of its FPC to be $.93, as compared with that of nonfat dry milk (NFDM), estimated at $1.57.

22. Rostad, personal communication, 1976.

23. Annan, personal correspondence, June 3, 1975.

24. *Ibid.*

Index